Azure Strategy and Implementation Guide

Third Edition

Up-to-date information for organizations new to Azure

Peter De Tender

Greg Leonardo

Jason Milgram

BIRMINGHAM - MUMBAI

Azure Strategy and Implementation Guide
Third Edition

Acquisition Editor: Ben Renow-Clarke
Acquisition Editor – Peer reviews: Suresh Jain
Content Development Editor: Dr. Ian Hough
Project Editor: Janice Gonsalves
Technical Editor: Saby Dsilva
Copy Editor: Safis Editing
Proofreader: Safis Editing
Indexer: Pratik Shirodkar
Production Designer: Sandip Tadge

First published: June 2020

Production reference: 2050620

Published by Packt Publishing Ltd.
Livery Place
35 Livery Street
Birmingham B3 2PB, UK.

ISBN 978-1-83898-668-1

www.packt.com

packt.com

Subscribe to our online digital library for full access to over 7,000 books and videos, as well as industry leading tools to help you plan your personal development and advance your career. For more information, please visit our website.

Why subscribe?

- Spend less time learning and more time coding with practical eBooks and Videos from over 4,000 industry professionals
- Learn better with Skill Plans built especially for you
- Get a free eBook or video every month
- Fully searchable for easy access to vital information
- Copy and paste, print, and bookmark content

Did you know that Packt offers eBook versions of every book published, with PDF and ePub files available? You can upgrade to the eBook version at www.Packt.com and as a print book customer, you are entitled to a discount on the eBook copy. Get in touch with us at customercare@packtpub.com for more details.

At www.Packt.com, you can also read a collection of free technical articles, sign up for a range of free newsletters, and receive exclusive discounts and offers on Packt books and eBooks.

Contributors

About the authors

Peter De Tender is a well-known Azure expert, a very passionate and dedicated technical trainer, who always manages to provide inspiring deep-technical workshops on the Azure platform, packed with demos and fun.

Before Peter joined the prestigious Azure Technical Trainer team within Microsoft, he held a similar position in his own company for the last 6 years. Now, he's continuing what he loves doing most, upskilling customers and partners on the wonderful world and capabilities of Azure.

Peter has been a **Microsoft Certified Trainer** (**MCT**) for +10 years, and has been a Microsoft MVP since 2013, initially on Windows IT Pro, but he moved to the Azure category in 2015.

Besides co-authoring this book, Peter has published other Azure-oriented material with Packt Publishing, Apress, and through self-publishing.

You can follow Peter on Twitter as @pdtit or @007ffflearning, or have a look at his website, http://www.007ffflearning.com, to stay up to date on his Azure adventures.

I would like to thank Ben Renow-Clarke from the Packt Publishing team for his trust and dedication to getting this book out, and having approached me as an author. What initially started as a small, quick writing gig expanded to a full authoring project of high quality and a massive amount of information. He also managed quality control to make sure all content was as up-to-date as possible throughout the writing process.

I also would like to thank my wife Els and my two daughters Kaylee and Kitana. You've had to miss me so many times in the last few years already, because I was away traveling to another location for another Azure workshop delivery somewhere in the world. And in the little bit of free time, you were still OK seeing me spending time on another side project. All because you know how it makes me happy, being able to inspire other people on this amazing platform called Azure. With this project being complete now, I can focus a bit more on all of you again. Thanks for the understanding, you are three amazing women!

Greg Leonardo is currently a Cloud Architect, helping organizations with cloud adoption and innovation. He has been working in the IT industry since his time in the military. He is a veteran, architect, teacher, speaker, and early adopter. He is a Certified Azure Solution Architect Expert and Microsoft Azure MVP. He has worked in many facets of IT throughout his career. He is President of TampaDev, a community meetup that runs #TampaCC, Azure User Group, Azure Medics, and various technology events throughout Tampa.

Greg has also authored the book *Hands-On Cloud Solutions with Azure* by Packt Publishing.

I would like to thank my wife Kate and two boys Maddux and Lucas for putting up with me while writing these chapters for this book.

Jason Milgram is the First Vice President and Cloud Solution Architect at the City National Bank of Florida headquartered in Miami, Florida. Previously he was VP of Platform Architecture & Engineering at Champion Solutions Group in Boca Raton, Florida.

Jason was educated at the University of Cincinnati and the Massachusetts Institute of Technology – Sloan School of Management. He was also a Sergeant in the US Army Reserve, serving from 1990 to 1998.

A Microsoft Azure MVP (2010-present), Jason has given over 100 Azure presentations and regularly writes articles on Azure topics.

About the reviewer

Steve Buchanan is a Director and Midwest Containers Services Lead on the Cloud Transformation/DevOps team with Avanade, the Microsoft arm of Accenture. He is an eight-time Microsoft MVP, and the author of six technical books. He has presented at tech events, including **Midwest Management Summit** (**MMS**), Microsoft Ignite, BITCon, Experts Live Europe, OSCON, and user groups.

Steve is focused on transforming the position of IT into a driver of digital transformation through ITSM, DevOps, and CloudOps. He stays active in the technical community and enjoys blogging about his adventures in IT at www.buchatech.com.

Credits

Adam Harbour

Ahmed Sabbour

Alessandro Segala

Enrico Fuiano

Guillermo Gomez

Joachim Hafner

Michael Leworthy

Ori Zohar

Shankar Sivadasan

Simon Schwingel

Table of Contents

1

Understanding the Azure Cloud

Introduction

Azure is a powerful platform that offers a multitude of services and capabilities for organizations of any size moving to a cloud strategy. Whether you are a start-up needing only the base infrastructure for running your company's website, or you are a multinational operating all over the world, any kind of organization today can start deploying and migrating workloads to Azure. The first step in taking advantage of the many capabilities offered by Azure is careful planning. It might be tempting to just dive in and go ahead with deploying Azure resources. However, it should be emphasized that building a cloud environment can require a similar level of planning and detail as building your own datacenter services!

This chapter is all about providing you with a clear overview of Azure's capabilities, its benefits, and how to get started in the correct way. It also touches on how to migrate existing workloads to the cloud and cover the tools that Azure provides to streamline and smoothen this process.

Next, this chapter introduces you to business-related questions where the cloud can help in innovation and digital transformation, as well as how to handle identity and security in the cloud. Last, it covers IT infrastructure-related questions on how to build your enterprise-ready networking, how to migrate file services, and what tools are available in Azure for monitoring and day-to-day operations.

Starting from the assessment phase, where you will receive guidance on what your organization's cloud readiness looks like, this chapter, and indeed the whole book, will help you understand the core strategy for app modernization with Azure.

Once you decide to start deploying and using resources in Azure, we will examine best practices around migration planning, migration tooling, and what processes Microsoft makes available for you to make this process as smooth as possible.

Aside from workload migration, you need an understanding of deploying and running Azure infrastructure services. Think of resources such as Azure networking, storage, and virtual machines, and how to manage these. Don't forget that Azure allows you to do more than just run your virtual datacenter in the cloud. More and more organizations are looking at public cloud solutions for hosting Platform as a Service-based workloads. This means you still run the business applications, but you are not deploying them on virtual machines anymore, and you are also not managing most of the infrastructure, such as networking and storage. Besides running infrastructure and platform services, you might also think of migrating your workloads to serverless and microservices solutions using containers. Having this flexibility regarding the environment you use in the public cloud, whether traditional infrastructure services, platform services, or containerized workloads, is really one of the core benefits of the cloud and a chief way in which Azure's technical innovation can support your business needs.

Furthermore, you must think about your organizational governance and compliance requirements. Even in your own datacenter, you don't just want anybody walking in and deploying new hardware, deploying new servers, expanding storage, and so on. The mindset around governance, security, and control remains, even if you start using cloud services. The good news is that Azure comes with an extensive list of governance and compliance capabilities, several even as a core part of the underlying platform. Others are offered as flexible, configurable services where you can take control.

Whereas the first few paragraphs in this introduction section mentioned the public cloud as an overall strategy for running your IT workloads, know that from here on we shift to Azure as Microsoft's public cloud solution. Before tackling more technical questions and aspects of what it takes to migrate and deploy your applications to Azure, let's talk about some scenarios where Azure can be of help in business innovation.

Business innovation with Azure

While cloud computing environments such as Azure have been around for 10 years now, the tipping point for cloud adoption by enterprises was seen around 2016, where for the first time an IDG survey found that over half of the IT environments of surveyed businesses were hosted in the cloud[1].

---

1   `https://bit.ly/2N8QVo4`

The "first generation" of cloud adoption was characterized primarily by the deployment of virtual datacenters. In this first generation, organizations deployed new or migrated existing virtual machine workloads into Azure for a variety of reasons: some migrated to the public cloud to save on datacenter-running costs, while others wanted to take advantage of the easier and faster method for the deployment of infrastructure. Other organizations looked to Azure to streamline their business processes, to use it as a test setup, or to use the cloud as an affordable disaster recovery solution. For others, the huge potential for performance and scale, as well as flexibility (especially during peak usage), were the core reasons for adopting Azure.

The "second generation" of the public cloud arrived when rather than purely seeing value in running and managing virtual machines, some organizations saw benefit and innovation in moving to platform services. This mainly removes the focus and dependency on virtual machines, networking, and storage, and switches to a new approach with the core focus being on the application itself. Since platform services don't require much infrastructure, they are easier to manage. There is also less time spent patching or maintaining servers, which typically results in the improved uptime of your applications as well.

A "third generation" is currently in its early stages, where organizations are adopting serverless and microservices, as well as using native cloud services to build cognitive solutions and artificial intelligence solutions. Azure makes it easy to deploy these kinds of back-end services, where less and less knowledge is needed to build the underlying infrastructure. For most of these services, there is not much, sometimes even nothing, to manage on the infrastructure side.

Azure, at its heart, is a public cloud platform, but there are a variety of different cloud models available in the industry today. Let's run through the main ones.

Public cloud, hybrid cloud, and multi-cloud models

At a high level, many organizations are looking into deploying or embracing one (or more) of the following cloud models.

Public cloud

This is the typical cloud platform offered by a service provider. Such service providers include Azure, Amazon Web Services (AWS), Google Cloud Platform (GCP), Rackspace, and Digital Ocean. In simple terms, the datacenter is managed by the vendor, and you consume the hosting part as a service.

Also, there is no dependency or integration with your existing on-premises datacenter. This is typically used by start-ups, **Small-Medium Business** (SMB) customers, or larger enterprises who want to build out a standalone environment outside of what they are running on-premises.

Hybrid cloud

In a hybrid cloud model, you are building an integration between your existing on-premises datacenter(s) and a public cloud environment. Most often, this is because you want to expand your datacenter capabilities, or you do not wish to perform a full migration to a public cloud-only model. Building a hybrid cloud typically starts with the physical network integration (in Azure offered by ExpressRoute or a site-to-site VPN), followed by deploying Infrastructure as a Service or Platform as a Service. Another aspect of the hybrid cloud is identity; Azure offers you Azure Active Directory as the identity solution. For hybrids, organizations synchronize (all or select) on-premises users and group objects from Active Directory domains to a single Azure Active Directory tenant. This allows for optimization in user and security management, offering users an easy but highly secure authentication procedure for cloud-running workloads.

Multi-cloud

More and more (enterprise) customers are looking at or currently using a multi-cloud strategy. Multi-cloud means using several public or hybrid clouds together. The benefit is using what is available. Imagine your business application relies on a service that is not available in your public cloud of choice, but might already be available in Azure. As long as you can integrate both worlds together in all aspects, such as security, supportability, skilled employees, and so on, there is no reason for not going in that direction. Looking at cost benefits could be another driver. Instead of running all workloads with the same public cloud vendor, it might be cost-effective to split workloads between different cloud vendors. Lastly, embracing things such as DevOps and Infrastructure as Code will also help you in adopting a multi-cloud strategy. Tools such as Jenkins, Terraform, Ansible, and several others provide REST APIs that can communicate with different cloud back ends. As such, your IT teams don't have to learn different cloud-specific templates, but rather can focus on the capabilities of the tooling instead of focusing on the cloud capabilities as such. At the same time, it should be mentioned a multi-cloud strategy also comes with several challenges. Supportability, mixed skillset requirements of your IT staff, and overall complexity because of the need to manage different environments are probably the most critical concerns to warn you about.

Azure public cloud architectures

Now you have a better understanding of the different cloud models, let's focus some more on public cloud architectures.

Infrastructure as a Service (IaaS)

Part of this first chapter is dedicated to migrating and running your business applications in an **Infrastructure as a Service** (**IaaS**) model, using similar concepts to those related to your on-premises datacenter, with virtual networking, virtual storage, and virtual machines as the main architectural building blocks.

However, that is not the only way that you can run your applications in Azure. As a segue to other chapters in this book, let me briefly describe where Azure can help in business innovation, or the digital transformation of your workloads, using other architectures besides virtual machines.

Platform as a Service options (App Service, SQL Database, Azure Container Instances, and Azure Kubernetes Service)

Platform as a Service (**PaaS**) refers to running your workloads in Azure without deploying virtual machines. You could, for example, run a web application in Azure App Service without deploying the underlying virtual machine first. This brings efficiency and optimization, since you have less operational management to worry about. Alternatively, you could use a service such as Azure SQL Database or Cosmos DB, allowing you to run the exact same flavor of your database solution, but again without having to take any virtual machines into consideration. As mentioned earlier, having this flexibility of architecture while also not needing to deploy and manage your own infrastructure dependencies anymore provides room for innovation. In addition, new capabilities are easier to adopt in a PaaS model than they are in an IaaS model, as the deployment of the service is much faster, thanks to the lack of operating system dependency and development language support, to name just a couple of reasons. PaaS can also be cost-effective, as most PaaS services are cheaper than their IaaS alternatives (for example, using Azure web apps is cheaper than deploying an Azure web virtual machine with the same performance characteristics).

Serverless (Functions, Cosmos DB, Logic Apps, and Cognitive Services)

Outside of these two standard concepts of IaaS and PaaS, you also have **serverless** and **microservices**. Looking at Azure serverless services, one option is to use Azure Functions, a serverless compute service that enables you to run code on demand without having to explicitly provision or manage infrastructure. You can use Azure Functions to run a script or piece of code in response to a variety of events. Another serverless service offered by Azure is Azure Logic Apps, a business workflow engine that provides connectors to more than 200 business applications, such as Dropbox, OneDrive, SAP, DocuSign, and Adobe. With Logic Apps, you can build a step-by-step workflow with logical intelligence to replace your current chain of complex, (mostly) virtual machine-based workloads.

Microservices is an application development approach in which a more complex application architecture, known as a monolithic application, gets divided into several smaller components. Each component has a single purpose, such as managing product ordering, order payment, or shipment follow-up, in a broader e-commerce platform solution. Microservices are a very popular strategy for migrating (legacy) applications to the public cloud. Microservices are also known as containers or containerized applications. At present, Docker (`http://www.docker.com`) is the standard container format for microservices and is fully supported by Azure.

Azure allows you to bring your containers into Azure and run them with **Azure Container Instances** (**ACI**) as a standalone container workload. Or, if you are looking for more advanced orchestration capabilities, you can also deploy **Azure Kubernetes Service** (**AKS**), providing a Kubernetes clustered environment as a service. Besides the advantage of legacy app support, containers are also very interesting because they are lightweight, running one specific task, and can easily be moved around across different environments. The same container image can run on a developer's Mac, be pushed into an Azure container registry, run on top of Azure Web App for Containers, run as an Azure container instance, or run in your own datacenter on top of RedHat OpenShift infrastructure or some other public cloud infrastructure. This is the main reason why containers are so popular today.

How can these different cloud architectures lead to business innovation? For a long time, businesses were struggling in optimizing IT and streamlining IT processes, searching for ways to align operations and developer teams—not always with huge success. However, by shifting from infrastructure (IaaS) to platform services or serverless (PaaS/SaaS), a huge part of this dependency is removed. The flexibility of using the public cloud and the different operational models it provides will definitely help in business innovation.

Outside of the technical IT-side of optimization, this could also lead to freeing up more resources and saving money that the business can then invest in new innovation. Let's look at some more reasons for moving to the cloud.

Strategizing for app modernization with Azure

Finding the right reasons for moving workloads to the cloud or deploying new workloads directly in the cloud is a vital part of the success of your migration projects. When you look at it through your system administrator's or developer's technical glasses, you will most probably find several good reasons for migrating. Systems can be deployed faster, operations and management become easier, and the cloud gives you scale, high availability, and cost efficiency. But maybe your upper management has a different opinion about it, as they just approved an extension of your physical datacenter. Or, maybe the industry you are in has several compliance regulations to follow, making the public cloud a challenging platform to approve.

Why move to Azure?

Let's try and be as open as possible about moving to the cloud, examining both the benefits and the risks. Know that these are mainly based on experience gained from working as an Azure cloud architect and trainer for several years. Also, keep in mind that what is a benefit for one organization might be a concern for another.

Cloud benefits

There are several common business challenges that make a move to the cloud worthwhile. Some typical scenarios where the cloud is beneficial are listed here:

- You are asked to reduce operational costs.
- Your applications are facing traffic increases and your in-house systems cannot scale to meet the demand.
- Your business requires faster deployments of systems and applications, in different regions in the world, to serve your customers.
- You have security regulations to follow that are hard to implement in your on-premises datacenters or are expensive.
- Your storage capacity is growing exponentially and it is hard to catch up, both technically and cost-wise.

- Your systems are becoming outdated, running no-longer-supported legacy operating systems and applications.
- The well-known CAPEX to OPEX (capital expenditure and operational expenditure): your company wants to shift to a consumption-based, pay-per-use model.
- You are a start-up, not having the financial means or certainty to deploy and maintain your own datacenter(s).

Potential challenges of cloud migration

While you might recognize several of the aforementioned examples in your own organization, thus giving you some good motivation to explore migration to the cloud, know that cloud migrations also come with potential challenges. While these are less and less seen as blocking factors, it is worthwhile to mention some of them:

- Your business is in a specific sector and you are not allowed to store data in the public cloud because of sensitivity or other compliance regulations.
- Your workloads suffer from latency when the datacenter is not close to where your users are.
- Some of your workloads are legacy; they cannot run in a cloud environment.
- You just made a huge investment in your own datacenter.
- You don't want vendor lock-in, and a multi-cloud strategy seems to have too much overhead.

Mapping business justifications and outcomes

If the preceding cloud benefits and potential challenges that come with cloud migrations have already got you thinking about cloud strategies, let's quickly run through a few other topics that you should keep in mind.

The cloud is not the cheapest solution for everybody

Switching or migrating from your own datacenter to the public cloud is a cost-effective solution, but that doesn't mean it is cheaper for everyone. Business analytics calculations might run faster because of cloud scale, which brings in business benefits, but the running costs might be as expensive (or even more expensive) than the costs associated with buying and running a similar workload on-premises. But it would take much longer before that infrastructure would be able to produce similar results.

To get a good view on the consumption costs of Azure resources, always start from the Azure Pricing page: `https://azure.microsoft.com/en-us/pricing/`.

No public cloud guarantees 100% high availability

This is one of my favorite conference topics to present on. Obviously, a public cloud platform is built with high availability in mind, but you need to be aware that there will not be availability 100% of the time. Looking at IaaS, you as the customer need to architect your high availability by using Azure Availability Sets or Availability Zones, for example. Moving to PaaS could be a solution for this, as it makes it more Microsoft's responsibility to make sure that your app services, data solutions, and so on are running on top of a high-availability platform. The good news is that Azure services have an overall **service level agreement (SLA)** of 99.9% for most services, with some services such as Azure SQL Database having a 99.99% SLA capability, or a 99.999% for Azure Cosmos DB.

All information related to Azure SLAs can be found at this link: `https://azure.microsoft.com/en-us/support/legal/sla/summary/`.

The lift and shift migration of virtual machines may not always give the best benefits

While a lift and shift migration of virtual machines to Azure is certainly a good step toward cloud migration, it is not always the most efficient one, from both a technical and also cost perspective. Maybe your systems are getting older, running legacy applications that might not run "better" in a cloud virtual machine. Or, maybe your on-premises systems were not correctly sized for the service they offer. Migrating such a workload to the cloud without changing the system characteristics might result in a huge cost increase for your Azure consumption.

Resources that can help in identifying your Azure **Return on Investment (ROI)** and **Total Cost of Ownership (TCO)** can be found here: `https://bit.ly/2R4wGc6`.

Containers are not always the best solution for cloud migration

We talked about containers a bit in the introduction. Containers are amazing, they truly are! Yet, they are not always the best platform to migrate your applications to, and they certainly shouldn't be the only driver for your cloud migration. Embrace containers as part of your overall cloud strategy, but don't see them as the endgame of your migration to the cloud.

For additional information on running containers on Azure and what services are available, have a look at this link: `https://bit.ly/2QFTf8h`.

What you learned in this section

You need to be aware of what the cloud can and cannot do for your organization. Decide whether any of the preceding maxims are true for your organization and your specific reasoning for moving to a public cloud scenario.

As you are reading this book, I am sure that you are interested in learning more about the technical side of deploying and running your business applications in Azure. That is exactly what this book is all about.

Any deployment or migration starts with knowing what you have before you are able to know where you are going. This phase is known as the **assessment** phase, so it makes sense for us to spend time on this phase in the next section.

Cloud migration approach

More than ever before, organizations are rethinking their datacenter strategy and heavily embracing and integrating the public cloud in their long-term IT strategy. Two key factors that are motivating this wave are cloud scalability and cutting down on capital investments. One of the starting points to make this public cloud migration successful is knowing what is supported in the public cloud from a workload perspective and knowing what your current IT landscape looks like. The better the match between these two, the easier your cloud migration will be.

Assessing your organization's cloud readiness

This first section is focused on assessing your on-premises datacenter(s) and determining what workloads, virtual machines, networking, storage, applications, and data solutions you have. You'll need to think about which ones you want to migrate to the public cloud and how to migrate them with the shortest downtime possible and lowest impact on the business. A successful migration relies on using tools to make the job easier. We'll look at several tools available from Microsoft that help in this process. We'll also go through identifying and estimating the cost aspect of the migration, and more specifically, what the running consumption cost could look like once the workload is running in Azure.

Assessment tooling

Helping its customers to perform an easy, or at least less complex, migration to Azure is one of Microsoft's key objectives. The easier (meaning less disruptive for the business) a migration is, the more success the cloud will see. Besides the excellent documentation on all things Azure at https://bit.ly/30kWG7B, you should have a look at the **Azure Migration Center**: https://bit.ly/384cHRN.

The Azure Migration Center breaks down the phases that you will need to run through in migrating to the cloud. The first of these is the **assessment** phase. This should always be the first phase of your broader migration project. Know what you have today to find out what you can run tomorrow.

Today, you can choose from the following free Microsoft assessment tools:

- **Azure Migrate**: Performs an assessment of virtual machine-based workloads
- **Azure Data Migration Assistant**: Runs an assessment for SQL Server databases, helping in migrating to Azure SQL virtual machines or Azure SQL databases
- **Azure Database Migration Service**: While not specifically an assessment tool, it helps in migrating from different data solutions on-premises to alternatives in Azure PaaS
- **App Service Migration Assistant**: Performs a scan of any application endpoint, supporting multiple languages and platforms (such as Java, .NET, Node.js, and PHP)

There is some overlap between assessment and migration for most of these tools, but let me spend some time on each, highlighting the core benefits and reasoning for checking them out.

Azure Migrate

Azure Migrate is the primary tool for running your workload migrations to Azure. It provides a full integration from assessment to migration and follow-up. The key strength of Azure Migrate is that it helps organizations in getting a detailed view on what their existing datacenter virtual machines look like from a characteristics perspective (CPU, memory, and disks), what operating system they are running, and most importantly, if they are supported for migration to Azure. Supporting both VMware and Hyper-V and up to 35,000 source machines (10,000 for Hyper-V environments), it is a viable tool for organizations of all sizes that want to validate migrating their virtual machines to Azure using a lift and shift approach.

Once Azure Migrate has a good view on the source virtual machine, it analyzes the gathered information and maps it with Azure. First of all, it identifies whether the virtual machine is compatible with the Azure virtual machine requirements (whether the operating system is supported, and so on), and it recommends an Azure virtual machine sizing. Lastly, it provides a monthly cost estimate for running that virtual machine as-is in Azure, presented in clear and easy-to-understand dashboard views (*Figure 1*).

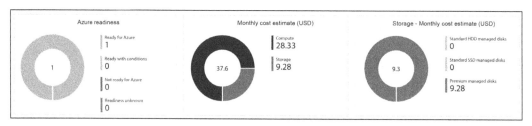

Figure 1: Azure Migrate – Assessment details

Another interesting feature is *dependency visualization*, where Azure Migrate will identify an application or other dependencies between virtual machines to smooth out the actual migration. Think of an application server having connectivity with a database server. It wouldn't help your migration if the application server was migrated over but the database server was not. Keep in mind that dependency visualization requires the installation of an agent, though (which might not be accepted by all organizations, especially when the decision to actually perform the migration to Azure has not been 100% confirmed).

VMware infrastructure assessment

In the case of an existing VMware infrastructure, where having vCenter is a dependency, you should deploy an *Azure Migrate appliance* that runs next to the existing virtual machines on VMware. To speed up this deployment, there is an OVA template available for download (around 15 GB), allowing you to deploy as an OVF template directly from within your VMware vSphere client and easing the installation of this appliance. This appliance performs a discovery, which can be scoped at vCenter datacenters, clusters, specific hosts, or select virtual machines. Once the discovery is complete, you will manage most of the assessment output back in the Azure Migrate project in the Azure portal, based on the metadata information sent from the appliance back to Azure Migrate.

Hyper-V infrastructure assessment

If you are using Hyper-V in your datacenter today, you can rely on Azure Migrate for the assessment and discovery. You start from downloading a Hyper-V VHD file (around 10 GB) and then configure it as a new Hyper-V virtual machine. Similar to the VMware scenario, this discovery appliance is responsible for gathering all necessary information from your on-premises infrastructure for the machines you define. It allows for a selective assessment based on Hyper-V hosts or individual virtual machines.

Physical and other cloud infrastructure assessment

With the latest version of Azure Migrate, Migrate not only assists in Hyper-V-based assessment but now also supports physical machines and other cloud providers (such as AWS and GCP) as a source. Similar to VMware or Hyper-V, the magic part is the *replication appliance*, starting from a Windows Server 2016 instance that you need to prepare.

Besides using the native Microsoft tools, know that the new Azure Migrate works directly with several vendors, offering third-party assessment and migration tools for different workloads. These are sometimes generic and sometimes support a single solution, but they are nicely integrated into the same Azure Migrate service today.

More information on what partners and tools are available can be found at the Azure Migration Center home page:

```
https://azure.microsoft.com/en-us/pricing/details/azure-migrate/
```

Azure Data Migration Assistant

If you have on-premises SQL Server-based data solutions running on physical or virtual servers, you can migrate them using the previously described approach, assuming the virtual machines are compatible with Azure virtual machine requirements. Though maybe you don't need to perform a lift and shift migration to Azure virtual machines, as your databases might qualify for running as Azure SQL instances. To find out, the easiest approach is to run an assessment using the Azure **Data Migration Assistant** (**DMA**) (*Figure 2*).

The assessment process could not be easier:

1. You run the tool in your on-premises environment, directly on the SQL Server machine you want to assess, or on a remote machine (for example, the SQL database administrator's workstation).

2. Once the tool is installed, you go through an assessment wizard, where you can specify the source (SQL Server) and target (Azure SQL Database).

3. After clicking on the **Start Assessment** option, the process begins analyzing your source database and usually completes doing so in minutes. The results are shown within the tool and can be exported as well. From an assessment perspective, it scans for two domains:

    ○ **SQL Server feature parity**: This is where you can review details about unsupported features and partially unsupported features, once your database runs in Azure SQL. One example I found out as a customer was the search index feature.

    ○ **Compatibility issues**: This is the second check that happens. If any issues should be detected, they will be listed here as a blocking factor for the migration:

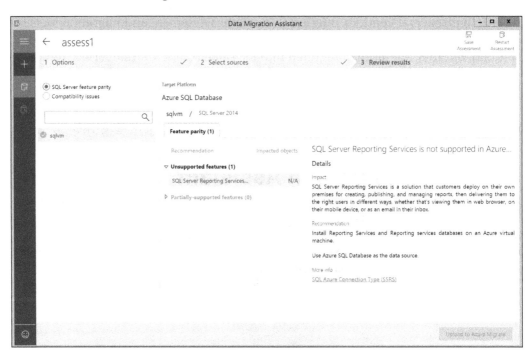

Figure 2: DMA – Assessment

Azure Database Migration Service

If, outside of SQL Server data solutions, you are running other databases in your on-premises datacenter, know that Azure offers another tool to assist you in getting them migrated over to Azure as well. This tool allows for smooth migrations from several different database sources to Azure data solutions. Typical uses cases here are:

- Migrating from PostgreSQL to Azure Database for PostgreSQL

- Migrating from Oracle Database to Azure Database for PostgreSQL

- Migrating from MongoDB to Azure Cosmos DB

- Migrating from MySQL to Azure Database for MySQL

So, typically, migration from an infrastructure virtual machine architecture to Azure PaaS, offers easy-to-follow wizards to help you in this process (*Figure 3*).

Note that this tool allows for both offline and online migration scenarios:

Figure 3: Azure Database Migration Service

App Service Migration Assistant

The App Service Migration Assistant is an updated version of the former Movemetothecloud.net tool. As the name suggests, the core functionality of this free Microsoft tool is helping in migrating your (web) applications to Azure Web Apps.

Besides performing the actual application migration, the other important capability is performing an assessment prior to the actual migration. The assessment starts by connecting to a public endpoint, or connecting to an internal web server, which gets scanned for several details of web server technologies to identify any migration issues, such as:

- Port bindings (since Azure only supports 80/443 for port bindings)
- Protocols (HTTP and HTTPS)
- Certificates (checks whether the site uses certificates, and self-signed or public CA)
- Dependencies in the `applicationhost.config` file
- Application pools
- Authentication type
- Connection strings

From a web server platform and language perspective, it supports a lot more than Windows Server IIS/.NET-based applications, such as:

- Ruby
- Node.js
- Java
- PHP

The outcome of the assessment and inventory is presented in a detailed web form, which can be exported for documentation purposes (*Figure 4*):

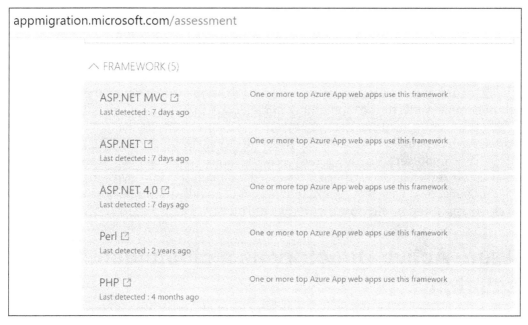

Figure 4: Azure App Service Migration assessment

Section summary

This section provided an introduction and description of the Microsoft application migration framework, by positioning and describing several Microsoft tools available to assist in the assessment process.

In the next section, the topic of identity and access control will be explained, as another important aspect to take into consideration before/during performing a cloud migration to Azure.

Identity and access control

Up until now, we have focused on the assessment and business justification side of your cloud migration projects, primarily from an application and data solutions perspective, but another important aspect to discuss is how you manage your identity and access control in the cloud. This is important for both end users and administrators.

Identity is the core component of all Azure cloud security: whenever any administrator wants to "do" something in the Azure platform, they need to authenticate and get authorization. No matter whether they are using the Azure portal, Azure command-line tools such as PowerShell or the Azure CLI, or using REST APIs. End users can also greatly benefit from Azure Active Directory. Solutions such as self-service password reset, Azure AD domain join for device management, conditional access, user risk, and many more will heavily optimize how users log on to cloud applications and how secure this log on is.

Azure Active Directory as a cloud identity solution

From an identity perspective, there is no way around Azure Active Directory. This cloud identity solution comes in different flavors:

- **Azure Active Directory**: The core identity component in Azure, offering cloud users, groups, applications, and service principal objects

- **Azure Active Directory Domain Services**: An emulated Active Directory service, offering Kerberos and NTLM, similar to your on-premises Active Directory domain controllers

- **Azure Active Directory B2B**: Business-to-business concept, whereby organizations can invite users from each other's Azure AD tenant

- **Azure Active Directory B2C**: Business-to-consumer concept, whereby organizations allow user authentication from social media identity providers (such as Facebook, Twitter, LinkedIn, and so on)

Besides the different flavors mentioned here, Azure Active Directory itself also comes in different editions:

Azure Active Directory Edition	Core Features and Capabilities
FREE EDITION	• Provides core identity services, storing users, groups, applications, and service principal objects • Can synchronize with your on-premises Active Directory using Azure AD Connect • Provides basic security reports
BASIC EDITION	• All features from the free edition + • Company branding • Application proxy toward on-premises web applications • Self-service password reset • Group management
PREMIUM P1 EDITION	• All features from the basic edition + • Self-service group management • On-premises password write-back • Two-way device write-back • Conditional access for optimized security
PREMIUM P2 EDITION	• All features from the Premium P1 edition + • Identity protection • Privileged identity management

Table 1: Azure Active Directory tiers

Just based on the rich feature set and advanced security features that come with it, any organization should consider *Azure AD Premium P1* for most of their cloud-enabled users, extended with *Azure AD Premium P2* for key users such as C-level management, administrators, security officers, and other key persons within the organization with high visibility.

Cloud authentication with Azure Active Directory

Most organizations already have an identity solution in place in their on-premises datacenter, often being Microsoft Active Directory. In this scenario, the recommended topology would be building out a hybrid identity architecture, starting from your Active Directory source environment. Azure AD Connect synchronizes the user and group objects (all or select ones based on filters you define). As such, a user account with the **User Principal Name** (**UPN**) peter@company.com from the on-premises Active Directory will authenticate with the same alias in Azure Active Directory.

However, there are three distinct authentication scenarios:

- Azure AD Password Hash Sync (PHS)
- Azure AD Federation using ADFS or third-party federation (ADFS)
- Azure AD Pass-through Authentication (PTA)

The easiest (and most recommended) approach is **Azure AD PTA**. In this scenario, your Active Directory objects are synchronized to Azure AD using AD Connect, including the domain's password hash. This allows users to log on to cloud apps using their Azure AD credentials, which are identical to the on-premises credentials.

Unfortunately, storing passwords (or the password hash) is a no-go for a lot of organizations, who want to keep control of the credentials from an on-premises perspective. In this scenario, you need to deploy a federation infrastructure, which can be **Active Directory Federation Service** (**ADFS**) or a non-Microsoft alternative (Okta is a popular one). While you still need to synchronize AD objects to Azure AD, the password is never stored in the cloud directory. Upon user authentication, Azure AD forwards the request to the ADFS infrastructure, which is typically running in the on-premises datacenter. ADFS sends the received credentials to Active Directory for validation. If these are accepted, the user can authenticate.

Whereas ADFS is the "typical" design to follow when deploying identity in a hybrid cloud model, it also comes with some drawbacks. ADFS servers run on-premises, which means there is a dependency on internet connectivity, as a highly available topology is needed to guarantee that users can always log on to cloud apps whenever needed. ADFS is also complex to manage, and your ADFS proxy server in the DMZ is public internet-facing all the time.

To accommodate the strengths and ease of use of password hash sync, together with the need to keep credentials management in the on-premises Active Directory, Microsoft came up with a third scenario, PTA. Again, you start by synchronizing users and groups with AD Connect. Next, instead of deploying a complex ADFS infrastructure, you deploy Passthrough Agents on your on-premises Active Directory Domain Controllers. These listen on port 443, but only to Azure AD services endpoints public IP addresses. Other requests will be denied. When a user logs on to Azure AD, the request gets passed on to the PTA agent, who sends along the credentials to the on-premises Active Directory, which is still responsible for validating the credentials.

Have a look at the following link for all details on Azure identity and access management documentation:

```
https://azure.microsoft.com/en-us/product-categories/identity/
```

Azure governance

Azure governance is a combination of different Azure services and capabilities, allowing for the management of all your Azure resources at scale and following control guidelines. Azure governance works across multiple subscriptions and across resource groups, and is based on a combination of Azure identity, **Role-Based Access Control** (**RBAC**), Azure policies, and management groups. You could extend the concept with Azure Resource Graph as well. Some customers also consider cost control as part of governance processes and best practices. If your organization has a **Security Operations Center** (**SOC**), this department will most probably take ownership of this process, or at least (should) be hugely involved in this.

Let me describe each of the different Azure services, allowing for governance.

Management groups

For a long time, an Azure subscription was considered the boundary of management and control. This allowed organizations to use multiple Azure subscriptions to "separate" resources from each other. Some organizations subscribed on a geographical level, some others used a dedicated subscription for a specific application workload, and others still separated based on dev/test and production.

This model changed recently, with the introduction of **management groups** (*Figure 5*). Where Azure Policy and Initiative were (and still are!) really great sources of governance control, they were linked to a single Azure subscription, which was hard to manage in larger Azure environments where admins wanted to replicate policy settings across multiple subscriptions. That's exactly what Azure management groups provide: a cross-subscription assignment of Azure Policy and Initiative.

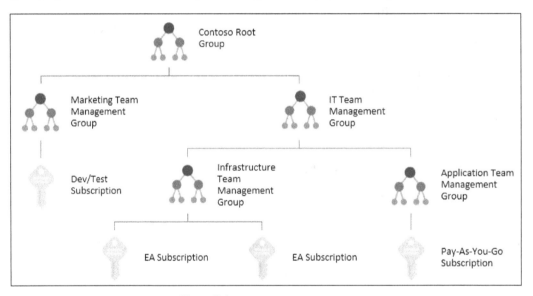

Figure 5: Azure management groups

Identity and role-based access control

Again, identity is key in a public cloud platform like Azure. The examples we saw earlier should already make it clear, but there is yet another example I can share: Azure itself heavily relies on RBAC to identify who can do what in the platform.

This "who" can be a user or group from your Azure Active Directory, a user from another Azure Active Directory tenant, or a registered application or service principal.

RBAC in Azure offers more than 75 different roles to choose from, and if you cannot find the specific role mapping for the particular need of your organization, you can create your own custom roles from Azure PowerShell as well.

Azure Policy

Another source of control is available through **Azure Policy**. This is a true governance management and control mechanism in Azure. As an organization, you define Azure policies: JSON files in which you specify what Azure resource requirements you want to enforce before the deployment of Azure resources can succeed. For example, there is forcing the usage of certain Azure regions because of compliance regulations, or allowing only certain Azure virtual machine sizes in your subscription to keep costs in control, or perhaps you might have certain naming standards you want to enforce for Azure resources, optimizing your asset management and CMDB regulations. One last example of something that a lot of companies find useful is enforcing the use of tags. A tag is like a label that can be attached to a Resource Group or individual resources, for example, a cost center or business unit. It is mainly thanks to these tags that an Azure billing administrator can get a clear view of what an Azure resource is used for, or at least to which business unit or cost center this resource belongs.

Azure policies can be grouped together into so-called **Azure policy initiatives**. This helps in enforcing several policies at once. After the Azure policies and policy initiatives are defined, they need to be assigned to a scope. This scope can be an Azure subscription, an Azure resource group, or individual Azure resources.

Azure Blueprints

Another mechanism available in Azure today for helping in governance control is **Blueprints**. Azure Blueprints (*Figure 6*) allows cloud architects and IT teams to define a structure of reusable, repeatable instructions for deployment and configuration, in compliance with company standards, regulations, and requirements.

Relying on a combination of roles, controls, and infrastructure as code, Azure Blueprints orchestrates the full deployment life cycle of Azure resources.

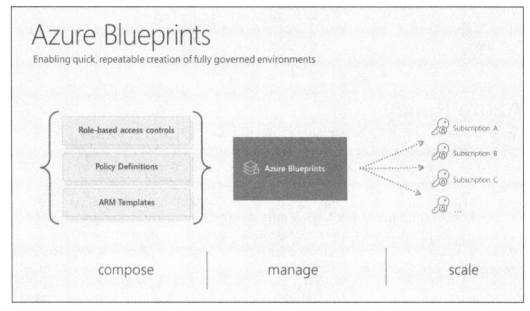

Figure 6: Azure Blueprints

Blueprints are based on artifacts, which are a collection of settings, parameters, Azure infrastructure as code deployment templates, and policy templates.

Naming standards

Another critical aspect of your migration strategy to a public cloud such as Azure is having a good understanding of the naming standards. Everything in Azure is based on Azure resources. Several of these use dynamic names that you cannot change. Other services are deployed in a fixed namespace domain (`azurecri.io` for Azure Container Registry, `blob.core.windows.net` for Azure Storage Account blobs, `azurefd.net` for Azure Front Door, and so on).

Next, several Azure resources have requirements (and limitations) around the usage of certain characters, capitals, and/or numeric values and complex characters.

There is good documentation available on this exact subject:

```
https://bit.ly/2FyxKjh
```

Resource groups

Another item I want to touch on as part of the pre-migration information is resource groups. While it's not that hard to understand what they do—they are groups of Azure resources—there is a lot of confusion around them, specifically regarding how to organize them, or how to organize your resources into them.

As a starting point, it is really up to your organization. Microsoft is not enforcing what resources should go where and how you want to organize your resource groups (with some exceptions). Some organizations have a resource group per workload; others define Resource Groups based on resource types (a Network Resource Group, a Storage Resource Group, and so on). An example of this approach can be seen in the diagram in *Figure 7*. This could help in allocating RBAC, keeping the same layered structure as their on-premises datacenter. Other organizations use geographical datacenter locations as a guideline (West-EuropeRG, East-USRG, and so on).

And touching on the topic of Resource Groups and Azure resource locations, specifying the location is a hard requirement for any Azure resource, as most resources on the platform are region-specific. Complexity arises when you have a Resource Group in one location containing resources in a different location. While technically fine, this might cause interruptions when the Azure region that the Resource Group is in is not reachable anymore. The resources would remain (for instance, a virtual machine would still be running), but you wouldn't be able to make any changes to the virtual machine (as the information metadata cannot be written to the Resource Group).

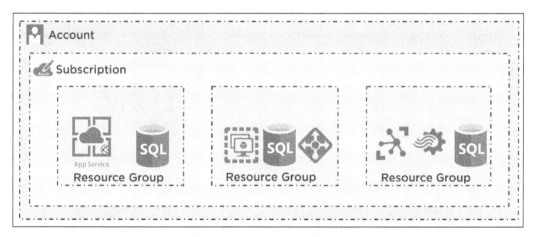

Figure 7: Azure Resource Groups

It should be clear now that identity and control are important topics to tackle before starting the actual migration (or deployment) of your business workloads on Azure.

By aligning the cloud architects with business needs, granting the correct roles and permissions, optimizing cloud security by using Azure identity features such as **Multi-Factor Authentication** (**MFA**), conditional access, privileged identity management, and Azure identity protection, you can dramatically optimize your security in the cloud. In most situations, that immediately means that you will also optimize the security of your on-premises datacenters, so this is a true hybrid-cloud benefit.

Azure Resource Graph

While not specifically built as a governance service, Azure Resource Graph can definitely help in getting a better view of the Azure resources an organization has deployed. Resource Graph is a service in Azure designed to provide a fast and easy-to-manage way to explore all resources within a single subscription, or even across multiple subscriptions.

Azure Resource Graph allows you to run filtering queries, narrowing the results of what you are looking for.

While Azure Resource Manager also allows you to gather filtered Azure resources, this is starting from the resource providers individually. If you want to get a view of Azure virtual networks, you would "call" the Network Resource Provider. Then, you would connect to the Virtual Machine Resource Provider to get information about your virtual machines.

Azure Resource Graph does this differently, and in a way that allows you to gather information across all those resources, without touching on each and every resource provider individually.

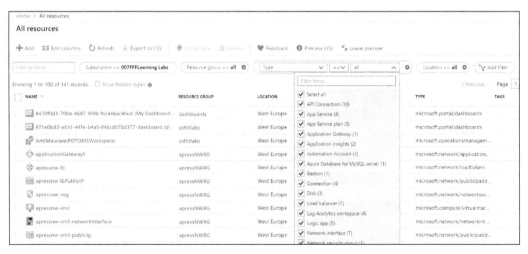

Figure 8: Azure Resource Graph

A result of this by using the Azure portal is shown in *Figure 8*. Besides the Azure portal, Resource Graph can also be used from Azure PowerShell and the Azure CLI, using the powerful and fast KUSTO query language.

Cost control and Cost Management

One last item that perfectly fits into the topic of Azure governance is Cost Management. Microsoft recently acquired Cloudyn, a multi-cloud cost reporting tool. The Cloudyn service enables any organization to pull up detailed dashboards, exposing cost consumption for any Azure resource or group of resources, based on resource type, region, or tags attached to the Azure resources itself.

Microsoft has now completely integrated the Cloudyn experience into the Azure portal, under a specific service called Cost Management, providing you with reporting dashboards (see *Figure 9*); there are several options for you to choose from.

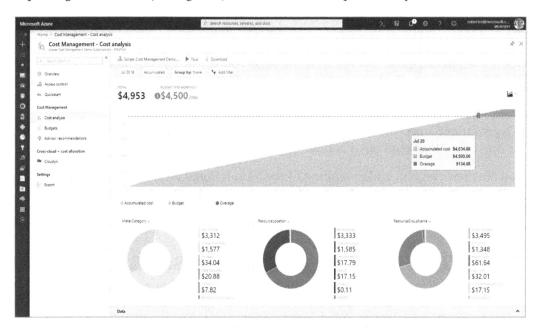

Figure 9: Azure Cost Management – cost analysis

Another recently released Cost Management feature is *Cost Budgets*. This is a soft setting, allowing you to define a ceiling of cost consumption for a certain Azure resource or resource group. Once the budget amount (or any percentage, such as 80%) is reached, Azure administrators can view a dashboard report of the results (see *Figure 10*) or receive an alert notification by email, for example.

Keep in mind that the budget feature is not stopping Azure consumption as such, and nor will the Azure resource be deleted, but it is at least a useful aid in cost governance.

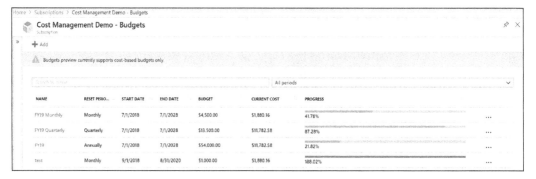

Figure 10: Cost Budget

Section summary

In this section, we guided you through several Azure governance services and capabilities you can deploy in your Azure subscriptions. Starting from Management Groups, allowing you to scope policies to multiple subscriptions at once, you learned about Azure policies and Azure Blueprints. We also talked about Azure identity as a governance mechanism, providing RBAC. In the last part, we covered the new service, Cost Management, as another governance instrument.

You now have a good understanding of the foundational layers of your cloud migration. Let's take a look at some of the migration tools and processes Microsoft has available today to help smooth this operation.

Migration tooling and processes

After having performed the assessment(s) of your source environment and laid out the path toward a solid cloud environment that takes identity, compliance, and governance into account, you are here to take the next step: running your migrations. We already briefly touched on several of these, but mainly from the assessment capabilities perspective that comes with them. This section walks you through the actual migration approaches. Several approaches are possible, and the one you choose will depend on different factors.

Manual migrations

The first option that comes to mind is running a manual migration. Starting from your existing source workload, you build a similar environment in Azure and copy over the data. Let's take a look at some examples.

Migrating VHD disks

If your source environment is Hyper-V and there is enough downtime allowed for your application, you could consider performing a manual migration by copying VHD disks into Azure Storage. Azure supports both Gen1 and Gen2 Hyper-V machines nowadays, with the maximum size of the VHD disk being 1,023 GB.

If your source environment is VMware, although there's a bit more work, it shouldn't be a blocking factor to copy the VM over to Azure. Tools such as Microsoft Virtual Machine Converter can help you in transforming the VMDK file into a VHD file format.

If you want to just migrate a single VHD and run an Azure virtual machine out of it, you don't have to generalize the VHD image; you can instead configure it as a specialized disk. If, however, you want to use this source VHD as a template disk for multiple Azure VM deployments, you should generalize the disk first. This is done using `sysprep` from within the source VM itself.

Once the disk is ready to be copied, use the `Add-azVHD` cmdlet in PowerShell to upload the VHD to an Azure Storage account. Other options you have available include using AzCopy and using the free Azure Storage Explorer tool.

Next, define an image from this uploaded VHD by using the `New-AzImageConfig` and `New-AzImage` PowerShell cmdlets.

Finally, when you have your image available, you can continue with deploying a new Azure virtual machine based on this source image.

Detailed step-by-step guidance on how to achieve this is documented here:

```
https://bit.ly/35zF8FK
```

Migrating SQL databases using bacpac

If your source database solution is an SQL Server database, you might be familiar with SQL Server's built-in backup solution, storing your database in a BACPAC file. This is the perfect way to migrate your database to Azure SQL, if downtime is allowed.

After you deploy a new Azure SQL Database instance in Azure (*Figure 11*), copy the bacpac file to Azure Storage. Head over to the import database option within Azure SQL Database and you're done! That really is how easy a migration can be.

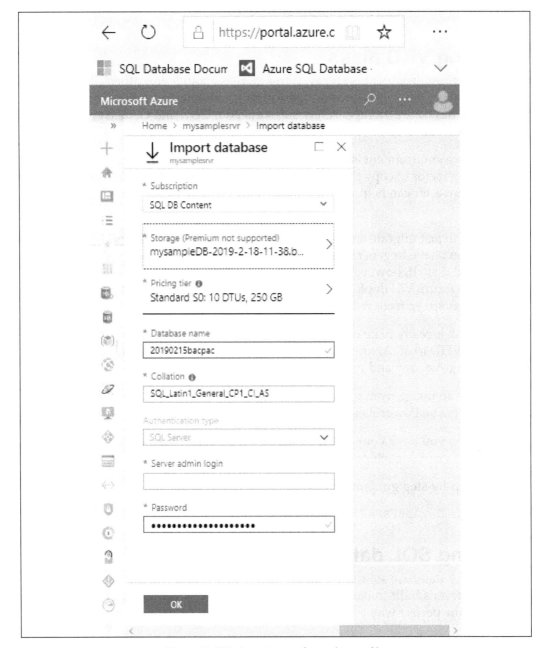

Figure 11: SQL Azure import from a bacpac file

Migrating websites to Azure Web Apps

The other common workload to migrate to Azure is a web application or a websites. Azure App Service's Web Apps supports both Windows web servers and Linux Apache web servers as underlying source environments, as well as a full list of development languages and frameworks (.NET, Java, Python, Node.js, Ruby, and more).

If you have your source code available, you could most probably push it directly into an Azure web app and run your site from Azure. Tools such as Visual Studio and Visual Studio Code provide this publish mechanism out of the box. If you have used the App Service Migration Assessment tool for an on-premises workload, know that you can also use the same tool to perform the actual web content migration step. *Figure 12* shows you a screenshot of what such migration process looks like from within the tool itself.

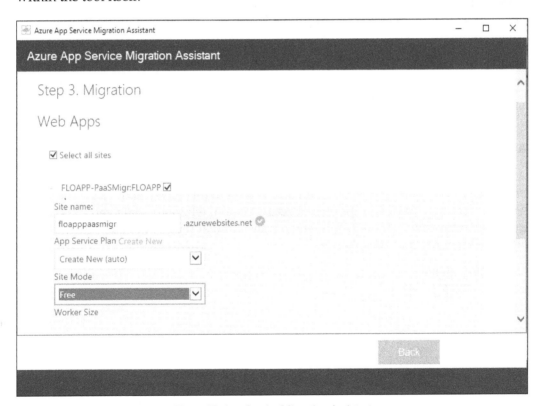

Figure 12: App Service Migration Assistant

Azure Migration Center

We already described the new Azure Migration Center in terms of its support and guidance in performing source workload assessments. But obviously, it can also guide you through the actual migration of these systems. Thanks to the latest updates, only published July 2019, Azure Migration Center has become the main place to check out when you need to migrate application or data solution systems to Azure. Therefore, Azure Migration Center has now become a central hub for starting, executing, and tracking your migration projects, whether you want to migrate to Azure virtual machines, Azure data solutions, or Azure App Service.

Azure Data Box

By now, it should be clear you have a full set of options to migrate your workloads to Azure, no matter whether the target environment is IaaS or PaaS. The aforementioned scenarios most probably cover about 80-90% of the migration scenarios.

There is one last scenario I would like to touch on here, which is the migration of larger volumes of data (think in hundreds of terabytes or even petabytes). These larger volumes of data are not the best candidate for manual migration, nor will you migrate those storage environments into Azure virtual machines using Azure Site Recovery. The good news is that there is another, more flexible way to do this: **Azure Data Box**.

Azure Data Box allows for offline data migration, based on a physical device you introduce in your datacenter. You offload the data to this physical device, which gets securely transported to the closest Azure region, where the data gets copied into your Azure environment. All data in transit and at rest is AES encrypted for advanced security and compliance.

Depending on the data volume size, you can choose between any of the three models shown in *Table 2*:

	**Azure Data Box:** • A rugged device, supporting up to 100 TB in capacity, offering standard storage area protocols, and integrating with standard data copy tools (such as Robocopy). This box offers 256-bit encryption.
	**Azure Data Box Disk:** • A rugged SSD disk with a SATA/USB interface, supporting up to 8 TB per disk. This disk offers 128-bit encryption.
	**Azure Data Box Heavy:** • This is for massive amounts of data, supporting up to 1 PB in data volume.

Table 2: Azure Data Box models for offline large data volume migration

Besides these **offline** data migration solutions, the Azure Data Box offering was extended with two additional **online** data migration solutions, Azure Data Box Edge and Azure Data Box Gateway, shown in *Table 3*:

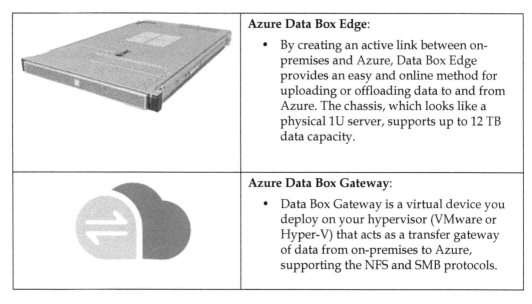

	Azure Data Box Edge: • By creating an active link between on-premises and Azure, Data Box Edge provides an easy and online method for uploading or offloading data to and from Azure. The chassis, which looks like a physical 1U server, supports up to 12 TB data capacity.
	Azure Data Box Gateway: • Data Box Gateway is a virtual device you deploy on your hypervisor (VMware or Hyper-V) that acts as a transfer gateway of data from on-premises to Azure, supporting the NFS and SMB protocols.

Table 3: Azure Data Box models for online data volume migration

Having discussed Azure Data Box, let's move on to talk about the deployment of Azure.

Deploying a greenfield Azure environment

Another migration strategy we should look at is deploying a greenfield environment in Azure, which means deploying a virtual datacenter from scratch. Instead of performing a lift and shift migration—as recommended in the previous scenarios—it could be beneficial to start with a totally new environment and only migrate the bare minimum (which would most probably be just data such as source code, file shares, and data solutions).

In order to make this approach a success, you must have a good understanding of how to architect and build such a greenfield environment, mapping it with your business requirements from a technical and non-technical perspective.

Obviously, the same is valid for the other migration scenarios described earlier.

For the remainder of this chapter, the assumption is that you will use IaaS, relying on the several virtual datacenter capabilities available in Azure.

Fundamentals of deploying Azure IaaS

As described earlier, IaaS refers to building a virtual datacenter. In the core, this refers to the following:

- Networking
- Storage
- Compute

Extended with overall security, governance, and monitoring capabilities.

Each of these architectural building blocks will be described in more detail in this section.

Networking

The core foundation of any datacenter, physical or virtual, is networking. This is where you define the IP address ranges and firewall communication settings, as well as define which virtual machines can connect with each other. Next, you also will outline how hybrid datacenter connectivity will be established between your on-premises datacenter(s) and the Azure datacenter(s).

From the ground up, you need to think about and deploy the following Azure services:

- Datacenter hybrid connectivity using Azure site-to-site VPN or ExpressRoute.
- Creating Azure virtual networks and corresponding subnets.
- Deploy firewall-like capabilities using Azure Network Security Groups, Application Security Groups, Azure Firewall, or third-party Network Virtual Appliances.
- Consider how you will perform virtual machine remote management. Regarding advanced security, just-in-time virtual machine access from Azure Security Center or the new Azure Bastion (public preview) might be good options. If neither of these, make sure your RDP/SSH sessions are behind a firewall scenario, never directly exposing your management host or virtual machines to the public internet.

- Just like in your own datacenter, Azure supports load balancing capabilities on its virtual network. You can choose from Azure Load Balancer, a layer-4 capable load balancer, supporting both TCP and UDP traffic on all ports. If you want to load balance web application protocols (HTTP/HTTPS), it might be interesting to deploy Azure Application Gateway, a layer-7 load balancing service, extendable with a **Web Application Firewall** (**WAF**) for advanced security and threat detection. The last option for load balancing is deploying a third-party load balancing appliance from trusted vendors such as Kemp, F5, or Barracuda.

- In the case of having multiple Azure datacenters for highly available workload scenarios, deploy Azure Traffic Manager or the new Azure Front Door service, allowing for detection and redirection across multiple Azure regions for failover and avoiding latency issues.

Besides these, running an Azure virtual network infrastructure — see *Figure 13* for an example diagram of what this network design could look like — is similar to what it's like in your own datacenters. Here are some of the core characteristics and capabilities:

- **Bring your own network**: This refers to the aspect of defining your Azure internal **virtual network** (**VNet**) IP ranges. Azure supports the standard class-A, class-B, and class-C IP addressing.

- **Public IP addresses**: Every Azure subscription comes with a default of five public IP addresses that can be allocated to Azure resources (firewall, load balancer, and virtual machine). Know that these IP addresses can be defined as dynamic or static. Also, keep in mind that it will never be a range of IP addresses, but rather standalone. It is also not possible to bring your own range of public IP addresses into an Azure datacenter.

- **Internal IP addresses**: When you deploy an Azure VNet and subnets, you are in control of the IP range (CIDR-based). Preferably, Azure allocates dynamic IP addresses to resources such as virtual machines (as opposed to the more traditional fixed IP addresses you allocate to servers in your own datacenter). In my opinion, the only servers requiring a fixed IP address in Azure would be software clusters such as SQL or Oracle, or when the virtual machine is running DNS services for the subnet it is deployed in.

- **Azure DNS**: Azure comes with a full operation DNS as a service that you can leverage for your own name resolution. When deploying an Azure VNet and subnet, it refers to this Azure DNS by default. You can change this setting to refer to your own DNS solution, however. This can be an in-Azure running virtual machine or an on-premises DNS solution (assuming you have hybrid connectivity in place).

- **IPv4 / IPv6**: Azure virtual networks support IPv4, but currently IPv6 support is in preview. More and more internal services, such as virtual machines and **Internet of Things** (IoT), are supporting IPv6 for its network connectivity, both inbound and outbound.

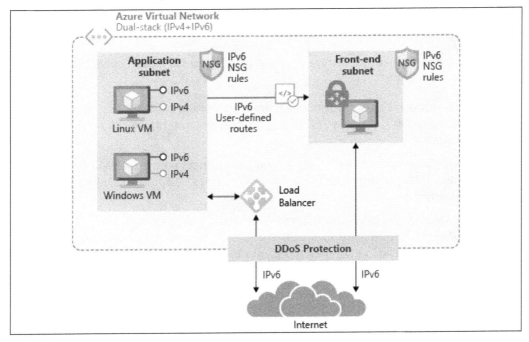

Figure 13: Azure virtual network architecture

Further details on Azure networking services, capabilities, and how to deploy and manage them can be found at the following link:

`https://docs.microsoft.com/en-us/azure/virtual-network/`

Storage

The next layer in the virtual datacenter architecture I will highlight is Azure Storage. Similar again to your on-premises datacenter scenario, Azure offers several different storage services:

- Azure Storage accounts for blobs, tables, queues, and files
- Azure managed disks for virtual machine disks
- Azure File Sync, allowing you to synchronize on-premises file shares to Azure
- Azure big data solutions

Let me describe the core characteristics of each of these services.

Azure Storage accounts

Azure Storage accounts are the easiest way to start consuming storage in Azure. An Azure Storage account is much like your on-premises NAS or SAN solution, on which you define volumes or shares.

When you deploy an Azure Storage account in an Azure region, it offers four different use cases:

- **Blob storage**: Probably the most common storage type. Blobs allow you to store larger datasets (VHD files, images, documents, log files, and so on) inside storage containers.

- **Files**: Azure file shares to which you can connect using the SMB file share protocol; this is supported from both Windows endpoints (SMB) a Linux (Mount).

- **Tables and queues**: These services are mainly supporting your application landscape. Tables are a quick alternative for storing data, whereas queues can be used for sending telemetry information.

Azure Storage accounts offer several options regarding high availability:

- **LRS – Local Redundant Storage**: All data is replicated three times within the same Azure datacenter building.

- **ZRS – Zone Redundant Storage**: All data is replicated three times across different Azure buildings in the same Azure region.

- **GRS – Geo Redundant Storage**: All data is replicated three times in the same Azure datacenter, and replicated another three times to a different Azure region (taking geo-compliance into account, to guarantee the data never leaves the regional boundaries).

- **GRS – Read Access**: Similar to GRS, but the replicated data in the other Azure region is stored in a read-only format. Only Microsoft can flip the switch to make this a writeable copy.

Azure managed disks

For a long time, **Azure Storage accounts** were the only storage option in Azure for building virtual disks for your virtual machines, dating back to Azure classic 10 years ago. While storage accounts were good, they also came with some limitations, performance and scalability being the most important ones.

That's where Microsoft released a new virtual disk storage architecture about 3 years back, called managed disks. In the scenario of **managed disks**, the Azure Storage fabric does most of the work for you. You don't have to worry about creating a storage account or performance or scalability issues; you just create the disks and you're done. Talking about performance, managed disks come with a full list of SKUs to choose from, offering everything from average (500 IOPS-P10) to high-performing disks (20.000 IOPS-P80), and there are also Ultra SSD disk types available, going up to 160.000 IOPS with this model.

It is important to note that the disk subsystem performance is also largely dependent on the actual VM size you allocate to the Azure virtual machine; the same goes for storage capacity, as not all Azure VMs support a large number of disks. Besides having larger disk volumes available from the operating system perspective, having a larger amount of disks available could also be used for configuring disk striping in an Azure virtual machine disk subsystem. This would also result in better IOPS performance.

Azure File Sync

If you have multiple file servers today, you most probably have a solution in place to keep them in sync. In the Windows Server world, this could be done using DFS — Distributed File System — which has been a core service since Windows Server 2012. When migrating applications to the cloud, you might also need to migrate the file share dependencies. One option would be deploying Azure virtual machine-based file servers for this. But that might be overkill, especially if those machines are only offering file share services. A valid alternative is deploying Azure File Sync.

By using Azure File Sync (see *Figure 14* for a sample architecture), you can centralize your file shares in Azure Files (as part of Azure Storage accounts), using them in the same way as your on-premises Windows file servers, but without the Windows Server layer in between. Starting from a Storage Sync Service in Azure, you create a Sync Group. Within this Sync Group, you configure registered servers. Once this server is registered, you deploy the Azure File Sync Agent on to it, which takes care of the synchronization process of your file shares.

Azure File Sync also provides storage tiering functionality, allowing you to save on storage costs when storing archiving data that you don't consult often but need to keep because of data compliance.

Data deduplication is another benefit for these cloud tiering-enabled volumes on Windows Server 2016 and 2019.

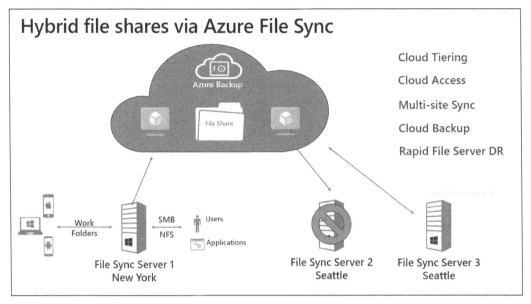

Figure 14: Azure File Sync

Compute

This brings us to the next logical layer in the virtual datacenter, deploying virtual machines. This is probably one of the most common use cases for the public cloud.

As mentioned in the introductory paragraphs at the beginning of this chapter, virtualization has dramatically changed how organizations are deploying and managing their IT infrastructure. Thanks to solutions such as VMware and Hyper-V, systems can be consolidated with a smaller physical server footprint, are easy to deploy, are easy to recover, and provide other benefits when it comes to moving them across environments (from dev/test to production, for example).

Most of the aspects and characteristics you know about from running virtual machines in your own can be mapped to running virtual machines in Azure. I often describe it as "just another datacenter." (However, obviously Azure is a lot more than that...)

Just like with the networking layer, let me provide you with an overview of the several benefits that come with deploying virtual machine workloads in Azure:

- Deploying Azure virtual machines allows for **agility and scale**, offering a multitude of administrative processes to do so. Leverage your expertise of PowerShell to deploy and manage virtual machines, just how you would deploy and manage them in your datacenter. Or, extend to Infrastructure as Code, allowing you to deploy virtual machines from Azure templates. This allows not only for Azure resources deployment but can be extended with configuration management tools such as PowerShell Desired State Configuration, Chef, Puppet, and more.

- Azure virtual machines come with a default SLA of 99.9%, for a single deployed virtual machine with premium disks. If your business applications require an even better SLA, deploy your virtual machines as part of a **virtual machine availability set**. This guarantees an SLA of 99.95%, whereby the different VMs in the availability set will never run on the same physical rack in the Azure datacenter. Or, deploy VMs in a **virtual machine availability zone**, leveling up the SLA to 99.99%. In this architecture, your VMs (two or more instances) will be spread across different Azure physical datacenter buildings in the same Azure region.

- You might have business requirements in place, forcing you to deploy virtual machines in different locations, hundreds or thousands of miles apart from each other. Or, maybe you want to run the applications as close to the customer/end user as possible, to avoid any latency issues. Azure can accommodate this exact scenario, as all Azure datacenter regions are interconnected with each other using **Microsoft Backbone cabling**. From a connectivity perspective, you can use Azure VNet peering or a site-to-site VPN to build out multi-region datacenters for **enterprise-ready high availability**.

- If you don't need real-time high availability, but rather are looking into a quick and solid disaster recovery scenario, know that Azure has a service baked into the platform known as **Azure Site Recovery**. Based on virtual machine disk state change replication, Azure VMs will be kept in sync across multiple Azure regions (keeping compliance and data sovereignty boundaries in place). In the case of a failure or disaster happening with one of the virtual machines, your IT admin can start a manual (or scripted) failover process, guaranteeing the uptime of your application workloads in the other datacenter region. The main benefit besides fast failover is cost saving. You only pay for the underlying storage, as long as the disaster recovery virtual machines are offline. During the disaster timeframe, you only pay for the actual consumption cost of the running VMs during the lifetime of the disaster scenario.

- Although the Azure datacenters are owned and managed by Microsoft, it doesn't mean they are only limited to running Windows Server operating systems and Microsoft Server applications. **60% of Azure virtual machines** are running a Linux operating system today.

- Many business applications, such as SAP, Oracle, and Citrix, are available in the **Azure Marketplace**, allowing for easier deployment, just like most other Azure virtual machine workloads. Starting from pre-configured images, any organization can deploy an enterprise application architecture in no time. Support for these third-party solution workloads is provided by Microsoft, together with the vendor.

- Azure offers **more than 125 different virtual machine sizes**, each having different characteristics and capabilities. Starting from a single CPU core virtual machine with 0.75 GB of memory, you can go up to virtual machines with 256 virtual cores and more than 430 GB memory. Next, you also have specific virtual machine families, supporting specific workloads (for example, the N-Series family is equipped with an Nvidia chipset for high-end graphical compute applications, the H-Series is recommended for **high-performance compute** (HPC) infrastructure, and you also have specific virtual machines sizes for SAP and SAP HANA).

For an overview of Azure virtual machines types and sizes, use this link to the Azure documentation:

`https://docs.microsoft.com/en-us/azure/virtual-machines/windows/sizes`

For a broader view on Azure virtual machine compute capabilities in Azure, use this link:

`https://azure.microsoft.com/en-us/product-categories/compute/`

The last Azure compute characteristic I want to highlight here is **Azure Confidential Computing**. In early May 2018, Microsoft Azure became the first cloud platform, enabling new data security capabilities. Mainly relying on technologies such as Intel SGX and **Virtualization-Based Security** (VBS), it offers **Trusted Execution Environments** (TEEs) in a public cloud. With more and more businesses moving their business-critical workloads to the cloud, security becomes even more crucial. Azure Confidential Computing aims at delivering top-notch security and protection for data in the cloud. The concept is based on the following key domains:

- **Hardware**: Intel SGX chipset from a hardware security perspective

- **Compute**: The Azure compute platform allows VM instances with TEE enabled

- **Services**: Secured platforms enable highly secured workloads such as blockchain
- **Research**: The Microsoft Research department is working closely with Azure PGs to continuously improve the capabilities of this trusted platform

Management of Azure infrastructure (and more)

The last part of the cloud migration and adoption process, regarding IaaS, is providing a solution that allows for enterprise-ready management of your Azure landscape. And where possible, this should be stretched to a hybrid management solution, especially during a longer migration period, where you have systems running in your on-premises datacenter but also in Azure.

Azure comes with a plethora of management, monitoring, and operations tools. Let me walk you through a few of these that you will most probably start using immediately on a day-to-day basis.

Azure Monitor

Azure Monitor provides a unified monitoring solution for Azure, offering a single place to go to extract insights from metrics such as CPU consumption, logs, including event and application logs, and any other telemetry information generated by Azure services. This can be extended to also monitor on-premises running workloads and solutions, resulting in a single, powerful monitoring tool.

Azure Monitor also offers advanced diagnostics and analytics, powered by machine learning. These are surfaced in Azure services such as Azure Monitor for virtual machines, Azure Monitor for Containers, Azure Advisor, and Azure Security Center, to name just a few. Each and every one of these canonical scenarios offers recommendations and insights into how your deployment of Azure resources is performing.

Monitoring in Azure typically falls into two categories: monitoring fundamentals, which are components that are available in the Azure platform automatically without having to enable any services, and scenario-specific monitoring, which includes services within the Azure platform that can be utilized for monitoring but will require additional configuration or may include additional costs. Examples of this would be Azure Monitor Log Analytics, which has a free tier and consumption-based pricing, or Azure Security Center, which also has a free tier and a paid tier. Azure Monitor is part of Microsoft Azure's overall monitoring solution. Azure Monitor helps you track performance, maintain security, and identify trends.

Azure Monitor enables you to consume telemetry to gain visibility into the performance and health of your workloads on Azure. The most important type of Azure telemetry data is the metrics (also called performance counters) emitted by most Azure resources. Azure Monitor provides several ways to configure and consume these metrics for monitoring and troubleshooting.

Azure Monitor information is based on two main log information types, metrics and logs. Metrics are numerical values, typically used for getting real-time feedback. Logs, on the other hand, contain much more detail and are typically used for retrieving or pinpointing correlations between events and activities. Logs would also be stored for a longer period of time, often depending on the compliance requirements of an organization.

Typical metrics for an Azure resource are published in the **Overview** section of an Azure resource (see *Figure 15* for an example related to Azure web apps).

Figure 15: Azure web app resource metrics

Retrieving more complex data from stored logs is done using Azure Monitor Log Analytics, where one needs to use Kusto query language-based queries. An example of what such a query looks like is shown in *Figure 16*.

Figure 16: Azure Monitor Log Analytics query example

See https://bit.ly/2QFzeyJ for more information and details.

Azure Monitor Log Analytics

Azure Monitor Log Analytics was previously treated as its own service in Azure, known as **Operations Management Suite (OMS)** Log Analytics. It is now considered a part of Azure Monitor and focuses on the storage and analysis of log data using its query language. Features that were considered part of Log Analytics, such as Windows and Linux agents for data collection, views to visualize existing data, and alerts to proactively notify you of issues, have not changed but are now considered part of Azure Monitor.

You require a log query to retrieve any data from Azure Monitor Log Analytics (*Figure 17*). Whether you're analyzing data in the portal, configuring an alert rule to be notified of a particular condition, or retrieving data using the Log Analytics API, you will use a query to specify the data you want.

You can create alerts based on your queries, and because you can include Azure metrics data in Log Analytics, you can even perform queries across your metrics and data stored in the Log Analytics service.

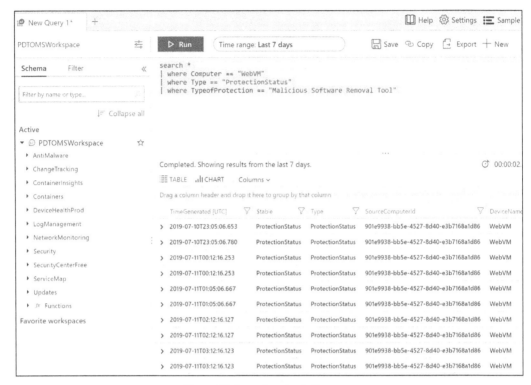

Figure 17: Azure Log Analytics query

Log sources are ingested from multiple services as well, just like metrics. In fact, metrics, along with activity and diagnostics logs, can serve as a data source for Log Analytics. Telemetry for virtual machines can be richer, as Log Analytics includes both textual and numeric data types, allowing for application logs, event logs, and additional performance metrics to be written to Log Analytics. The same is true for application data for custom applications. Applications can be instrumented with Application Insights to provide deep insights into application performance and health. Azure Security Center also leverages Log Analytics as a part of its analysis of the virtual machines in your subscriptions. When onboarding a VM to Security Center, you're in fact onboarding that machine to a Log Analytics workspace where its telemetry is stored for analysis by Security Center. There are also marketplace solutions that extend a Log Analytics workspace, bring additional data points into the service for a query, as well as new visualizations based on those data points.

Finally, you can interact with and respond to events based on Log Analytics data with services such as Azure Automation, where you can have an event trigger a runbook or even a webhook in an Azure function.

Azure Security Center

Azure Security Center (*Figure 18*) provides unified security management and advanced threat protection across hybrid cloud workloads. Many organizations are moving workloads to the cloud or deploying new workloads in the cloud to optimize their security posture.

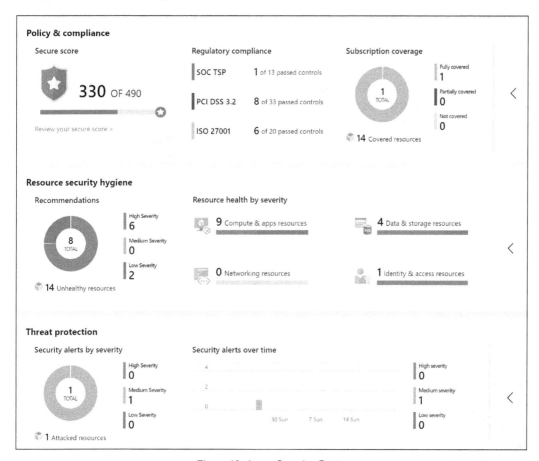

Figure 18: Azure Security Center

Azure Security Center offers powerful capabilities across three main areas:

1. **Cloud Security Posture Management**: Azure Security Center provides you with a top-down security posture view across your Azure environment, enabling you to monitor and improve your security posture using the Azure Secure Score. Security Center can help you identify and perform security best practices and hardening tasks and implement them across your machines, data services, and apps. This includes managing and enforcing your security policies and making sure your Azure virtual machines, non-Azure servers, and Azure PaaS services are compliant. With newly added IoT capabilities, you can also reduce the attack surface for your Azure IoT solution and fix issues before they can be exploited. In addition to providing full visibility into the security posture of your environment, ASC also provides visibility into the compliance state of your Azure environment against common regulatory standards.

2. **Cloud Workload Protection**: Azure Security Center's threat protection enables you to detect and prevent threats at the IaaS layer as well as in PaaS resources in Azure such as IoT and App Service, and finally with on-premises virtual machines. Key features of Azure Security Center threat protection include config monitoring, server EDR, application control, and network segmentation. Azure Security Center also supports container and serverless workloads.

3. **Data Security**: Azure Security Center includes capabilities that identify breaches and anomalous activities against your SQL databases, Data Warehouse instances, and storage accounts, with support for other services on the way. In addition, Security Center helps you perform the automatic classification of your data in Azure SQL Database.

Security is part of all layers of the public cloud environment. The good news is that Azure Security Center has the capabilities to tackle and report on each of these layers in an Azure environment, as well as in a hybrid cloud scenario.

Azure Sentinel

While Azure Security Center is mainly used as a reactive operations tool, many organizations are struggling to switch to a reactive approach. This is where **Azure Sentinel** can be of value. Azure Sentinel is a **Security Information and Event Management (SIEM)** tool in the cloud. Given the huge amounts of security-related data out there, it is important to get a clear view and focus on the key concerns and how to fix them. A lot of different Microsoft services, such as Azure, Office 365, and also non-Microsoft services, can report information back to Sentinel.

All this information goes through a massive data analytics and machine learning engine, helping to identify threats. Also, it comes with more than 100 built-in rules, and you can configure your own alert rules.

When one or more incidents (cases) occur, they will be centralized to allow for further investigation and handling. With a powerful graph view, it becomes easier to present and detect the correlation between stand-alone attacks and vulnerabilities. And all of this is available cross-platform, whether checking Azure resources, on-premises, or hybrid.

Besides the dashboarding views and detections, Azure Sentinel can also help in remediation. Using Security Playbooks—backed by Azure Logic Apps—it doesn't take more than a few clicks to build a step-by-step logical approach of action-taking.

Lastly, a huge community of Microsoft and non-Microsoft security experts and organizations helps to maintain and optimize security detections, centralized in the Sentinel GitHub repo, `https://bit.ly/302Yfqq`, resulting in an even more powerful security service.

Azure Network Watcher

A core activity as part of managing your own datacenter is getting a clear view of your network traffic, not only from a security perspective but also regarding bandwidth, latency, and suchlike. Vendor tools such as Wireshark and Fiddler have been part of any IT admin's troubleshooting toolbox for years. While these tools are still useful in an Azure environment, they do not always give the same rich, detailed experience. The main reason for this is because you don't own and control the Azure network stack as you do in your on-premises datacenter. There are no firewalls or switches that you can connect a serial console cable onto in order to capture network traffic or read out syslog information.

And that is where Azure Network Watcher (*Figure 19* gives a dashboard view of Network Watcher integration with Azure Monitor) comes in. Bringing in a lot of typical network trace features, it is a must-use tool if you take Azure network monitoring seriously:

- **IP flow verify**: Allows you to get a clear view on the matter if a packet is allowed or denied to or from a virtual machine, using five-tuple information

- **Next hop**: Provides the next hop from the target virtual machine to the destination IP address

- **Effective security rules**: Shows the effective rules, based on different **Network Security Group** (**NSG**) rules, configured on different levels (NIC, subnet, and so on), thereby making it easy to troubleshoot and see the effective reason why traffic is allowed/denied

- **VPN troubleshoot**: Diagnoses the health of site-to-site VPN gateways and ExpressRoute, capturing details in a diagnostic log file
- **Packet capture**: Similar to Wireshark, runs packet captures to analyze all the details of the network stack

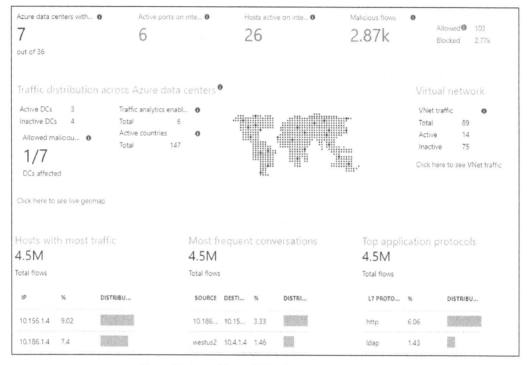

Figure 19: Azure Network Watcher connection monitor

Azure Service Health

Sometimes you face an issue or downtime with one of your running Azure workloads. But this is not always caused by your administrators. Like any other datacenter, Azure undergoes continuous updates and requires patching of its infrastructure. Physical components fail all the time, especially if you consider the amount of physical servers, storage, networking, and racks that are running inside each and every Azure region. To help you in troubleshooting, as well as to legally identify you about any issues being faced by the PaaS provider, Azure gives you Azure Service Health (*Figure 20*). From here, you can get a real-time view of the overall uptime status of any Azure datacenter you have services running with, as well as historic views. If you detect an issue with a workload or service whose configuration has not been changed by yourself, it should be a reflex to open Azure Service Health and validate that all is fine on the Azure physical stack.

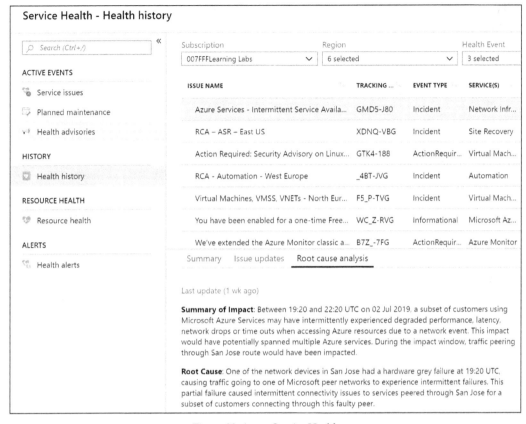

Figure 20: Azure Service Health

Azure Advisor

Azure Advisor is yet another Azure monitoring tool, and it is a great one. First of all, it's free, but that's not the main reason why any Azure customer should enable and use it in their subscriptions.

Azure Advisor provides insights and recommendations in four domains:

- High availability
- Security
- Performance
- Cost

All of this is presented in a nice Azure dashboard (*Figure 21*), allowing you to drill down to see more details regarding the recommendations.

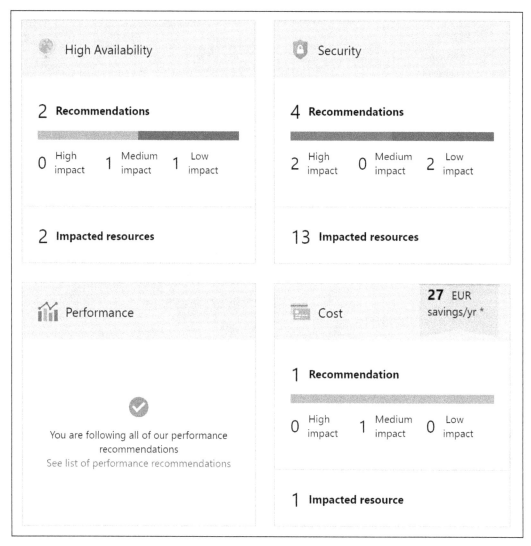

Figure 21: Azure Advisor

The core idea of Azure Advisor is getting guidance on using Azure best practices, helping you optimize running workloads and services in Azure. Based on machine learning in the back end and relying on telemetry information and configuration settings from your actual Azure environments, it will provide recommendations for optimization.

While it would be possible to go after that information yourself, it is probably a lot more time-consuming and cumbersome to find than what Azure Advisor presents nicely and almost in real time, without any hassle or agent needing to be deployed.

Azure Monitor Application Insights

Azure Monitor Application Insights is yet another Azure monitoring tool, with a core focus on monitoring your (web) application landscapes, no matter where they are running. And that is immediately one of the main reasons why you should have a look at it. Although running in Azure, the to-be monitored web apps don't need to run in Azure App Service. Next, it is a very detailed tool, capturing most of the insights a developer is looking for when running web applications. It detects information regarding web app performance, but also helps in analyzing web app traffic. From a language perspective, most popular environments and development languages are supported (.NET, Java, Node.js, Python, Ruby, and more).

You install a small instrumentation package in your application and set up an Application Insights resource in the Microsoft Azure portal. The instrumentation monitors your app and sends telemetry data to Azure Monitor. (The application can run anywhere—it doesn't have to be hosted in Azure.)

Information that Azure Monitor Application Insights is capable of tracking and presenting includes:

- Web app response times and failed requests
- Exceptions in traffic or usage
- Page views and load performance
- Custom metrics, if configured

It allows you to work with four domains of web app hosting:

- Monitoring (availability, performance, and integration with other services such as databases)
- Detecting and diagnosing issues (HTTP errors, providing traces, and so on)
- Synthetic transactions (running web tests)
- Telemetry information

The different sections that are monitored by Azure Monitor Application Insights can be presented in clear and customizable dashboards (see *Figure 22* for an example), which can be shared among your Azure administrators.

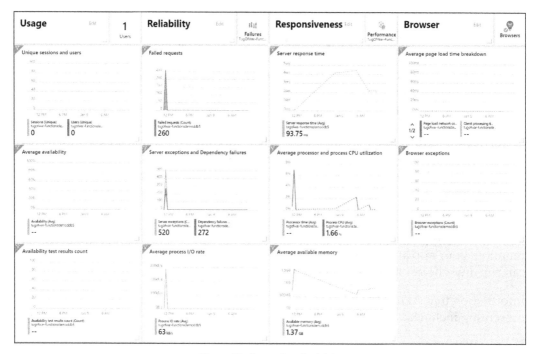

Figure 22: Azure App Insights

Chapter summary

In this first chapter, we outlined the Microsoft best practices around Azure migrations, starting from assessments and what tools Microsoft provides to help in this phase. Next, we discussed Azure Migrate and the newer Azure Migration Center. We looked at how it provides the necessary and useful tools for performing actual workload migrations from on-premises to Azure, covering different architectures. Next, we touched on the different foundational aspects of running a virtual datacenter in Azure, highlighting identity and control, enterprise-ready networking, Azure Storage capabilities, and what Azure virtual machine architectures are available today.

We provided information on business innovation and the digital transformation of traditional workloads with virtual machines, and what other Azure capabilities and services are available to host your business-critical workloads using PaaS, serverless, and microservices.

We also shared insights on Azure monitoring and operations, including how to efficiently manage your cloud environment using core Azure built-in tools.

While most organizations have been using a multitude of similar tools to manage and operate on-premises datacenters, there is no need to worry about additional complexity because of more Azure services and tools being added. Depending on the cloud strategy, an organization can initially start with using the Azure-provided operation services for monitoring and operating Azure workloads only. Yet, know that most of the monitoring services referenced here also extend to your on-premises datacenter, and most often hybrid scenarios as well. So, it might be a good approach to learn how to integrate these Azure tools in your overall IT landscape and leverage their true power. Don't get stuck in an Azure-only mindset.

If you would like to try Azure and make use of some of the features mentioned in this guide, you can sign up for a free account at the following link: https://azure.microsoft.com/free/services/virtual-machines/.

2
Architecture Choices and Design Principles

As organizations work to modernize their applications, either for themselves or their clients, they aim to maneuver their apps toward scalability, resiliency, and high availability. The cloud and mobile devices are changing the way in which organizations approach application design. We are seeing large monolithic applications being replaced by smaller decomposed or decentralized services. These services provide communication through microservice APIs, or asynchronous messaging or eventing. This shift has created new hurdles for organizations to overcome, such as parallelism, asynchronous operations, and distributing application state. There are also core considerations to keep in mind, such as designing for failure, or scaling while embracing the automation of management and deployment.

This chapter shows an approach to architecting cloud solutions covering a variety of technologies and topics. We will look at some popular technologies within the cloud and review best practices and some pricing considerations, but first let's examine some fundamentals of cloud application development before we dive into the technologies.

Application fundamentals for the cloud

We'll begin with some key takeaways in cloud versus on-premise development practices. In the cloud, there are many ways to solve a problem, so work at keeping it simple – try not to engineer your way to complexity. Keep in mind the basics behind application architecture, shown in *Figure 1*, and work at solving the layers as simply as you possibly can.

Everyone in their journey to the cloud has been bitten by the complexity bug and has learned that lesson:

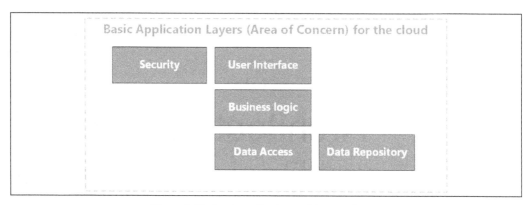

Figure 1: The basic application architecture layers

Tackling these layers and defining them as early as possible is an essential part of the cloud architecture and will enable each part of the organization to contribute to the application as a whole, meaning security, development, operations, and testing get an early seat at the table. The biggest part of being a successful cloud architect is learning to empower those around you to be successful in their roles and to feel that they are part of the process. Note that these layers do not need to be in the same project and can be broken out and decoupled, but, at the end of the day, applications have distinct boundaries. Remember not to comingle the layers, for example, putting business logic in your UI or data repository. Architectural guardrails are a key part of cloud architecture, and we will cover project structures more deeply in *Chapter 3, Azure DevOps*.

Keeping the basic application layers in mind will lead to overall cloud governance that will help place guardrails and empower architects to choose the right resources to deliver the best solutions. Cloud corporate policies should include definitions around the business risk that will help define policy and compliance and create the process to monitor for adherence to these defined policies. Your cloud corporate policy should follow the following tenets:

- It should document what has been identified and understood as the corporate risk
- This risk should be translated into policy statements
- These statements should be tested and monitored for adherence

As organizations begin their journey to the cloud, it is important to gain a good overall understanding of the architectural styles, design, technology choices, and the use of the five pillars of software quality (resiliency, security, availability, scalability, and management); as a side note, I find it useful to keep a link to `https://docs.microsoft.com/` to review any Microsoft architectural updates, with respect to native cloud application development. Checking your new architecture against the five pillars will help your applications succeed. Along with sound architectural practice, it is also important to understand the areas of responsibility in the cloud. Troubleshooting in the cloud can hurt at times without this understanding of the areas of responsibility because you may not have access to the underlying logs, or the logs might be streamed and difficult to follow. This makes it extremely important to understand the areas of responsibility and how the five pillars are affected by them. Let's take a quick look at these responsibilities in *Figure 2*:

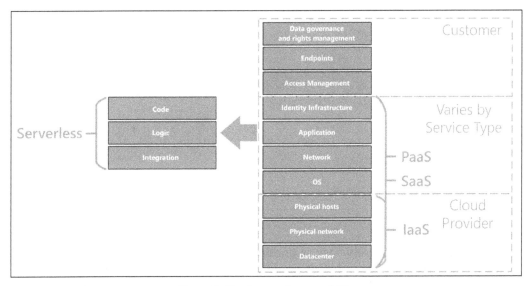

Figure 2: Cloud areas of responsibility

Figure 2 shows a cloud responsibility matrix. While in an on-premise setup you or your organization are responsible for all the layers; in a cloud setup, responsibility is shared with the cloud provider. When you move from on-premise or dedicated hardware in a datacenter, there are some new boundaries for where responsibilities lie. This shift in responsibilities can create undue fear within an organization's operations team, but they need to understand that these changes will help enhance their operation at the end of the day. This change in responsibility also creates different financial responsibilities, as the autonomy of cost now reflects computation time or storage space, and is more of a utility cost than a sunk or cap cost.

You should always try to keep in mind the basics of what it means to move from on-premise to the cloud. *Figure 3* shows how a simple application stored on a **virtual machine** (**VM**) maps between on-premise and Azure:

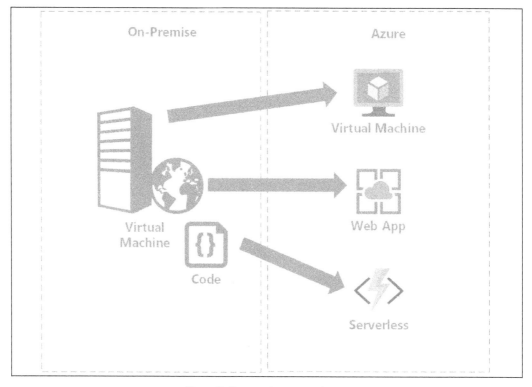

Figure 3: On-premise versus Azure

As you can see, it is relatively easy to understand a VM-to-VM mapping – moving the VM as a whole to the cloud. This model, referred to as "lift and shift," is the easiest but is also the most costly. This is the fastest way to migrate to the cloud without any code modifications. This model relies on Azure VNets, virtual network appliances, and domain services.

You can choose to move only your applications from IIS to a web app, which is a little more complex in nature because you may need to modify your code to complete this migration. This type of move helps your cost model and lines of responsibility because you only need to worry about the application and not the underlying resources needed for that application. For example, applications that are housed within IIS or task manager can be moved out to Azure Web Apps or WebJobs by changing your deployment endpoint.

You have the ability to take this a bit further by using a serverless model. Here, you can move just the underlying code class structure into a serverless model with Azure Functions. This takes a bit more testing and modification, but is the cheapest model from a consumption and support perspective. You need to consider how the various parts of your business will map from on-premise to Azure and keep that in mind when it comes to picking your architecture.

Now let's look at some of the challenges and considerations that you will need to keep in mind as we cover the various architectural options in this chapter.

- Ensure that components are designed to be consistent, manageable, and reusable, with an eye on the implementation aspects of administration and deployment. Decisions and guardrails put in place during the design phase will have a great impact on the overall outcome and on your ability to pivot your design when needed.

- Applications in the cloud exist in the public domain outside of your organization, so security becomes a fundamental principle in order to restrict access and protect data. It is good to understand how to delegate authentication or to federate identities to external providers. For example, you might use a storage account that utilizes a "valet key," which is a token/ key to restrict access to important resources. Using a gateway or gatekeeper to isolate your services and applications helps protect and broker requests with the clients consuming them.

- One important thing that you will require in the cloud, as with any shared model, is resiliency – being able to handle and recover from issues that occur. You need to be able to detect issues and to react to them quickly and efficiently. Placing "bulkheads" or isolating services from each other to ensure failures don't propagate is an essential tactic and should be used with a "circuit breaker" that further isolates services. The "circuit breaker" pattern assists with service isolation until the service is back in a healthy state, just like a circuit breaker in your house. You also want to take into account how to roll back from a fail state in the disconnected system. Look into "compensating transactions" that enable you to step back through a process until you can proceed with a consistent operation. You also want to use health-check monitoring of your endpoint and ensure you have a retry process that can work with these types of failures.

- Availability is a key foundation in the cloud, enabling you to spread workloads and to throttle the consumption of your resources, and it is usually measured in uptime, just like it is from an on-premise perspective. The biggest difference between the cloud and on-premise is the resource-on-demand model that the cloud provides, where resources can be autoscaled up and down with little to no intervention.

What this means is that to scale an on-premise solution, you need the virtual space or new machines to scale. This also includes scaling the whole machine and not the smaller components and requires adjustment to the server farm or network appliances to add the new machines.

- Another key element in the cloud is how you handle data management, as data is typically hosted in multiple locations on multiple servers. You must ensure that the consistency of your data is well maintained and synchronized across multiple locations. Making sure that you have a caching strategy for static content or frequently requested data elements is essential, as they can play a major role in application performance.

- To maintain performance and responsiveness you should have a process for managing variations or changes in workload processing and for dealing with peak usage times and system loads. One of the great benefits of moving to the cloud is how this scaling and resource allocation is mostly handled automatically. When designing a system architecture, you should as much as possible make use of resource types that support this automatic scaling.

- With all this complexity, the ability to manage resources and monitoring jumps to the forefront of your journey to Azure. While resource management and monitoring were important for on-premise models, they are critical for the cloud where the boundaries of responsibilities change, as shown in *Figure 2*. These changes shift your focus away from what hardware your application is running on, and on to areas that are important to your application.

- Messaging in the cloud is used to decouple service dependency in the same way that service buses were used in on-premise, but this disconnected asynchronous structure has its own challenges, such as ordering, message management, and idempotency.

- One of the most important pieces to get right when implementing a cloud strategy is to put a good naming convention in place. This is important for troubleshooting and making human-understandable systems. It is also difficult and time-consuming to make changes to file naming systems further down the line, so it's worth getting it right from the outset. Microsoft's best-practice recommendations can be found at `https://bit.ly/37Kj0tI`.

We should also recognize that not everyone wants or needs to move their entire organizational workload to the cloud, so we will also point out hybrid models during this chapter that mix cloud and on-premise solutions. *Figure 4* shows a quick example of a hybrid scenario:

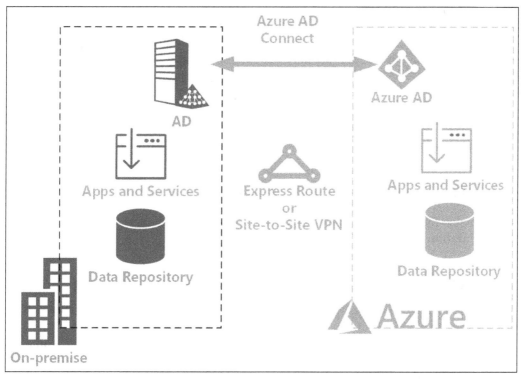

Figure 4: Simple cloud hybrid model

As you can see, the connection can either be through a simple site-to-site or ExpressRoute VPN to connect the on-premise network to Azure, and Azure AD Connect is used to sync the network AD with Azure AD so that security and **Single Sign-On** (**SSO**) are persisted from on-premise to the cloud. This would be the same basic approach to forming a cloud-to-cloud or on-premise-to-cloud setup. There will be differences between the cloud-to-cloud technologies you use, and these will need to be settled using VMs or containers to ensure code compliance as well as compatibility. While most cloud providers share some of the same functionality, Azure excels in providing many different native services to help solve issues. Containers, however, provide a better cross-platform experience without the need for reengineering your code base; we will cover this a little later in this chapter.

One other key thing to point out before we dive too far into architecture is that DevOps has great synergy with your journey to the cloud. This synergy is around consistent deployment and **Desired State Configuration** (**DSC**) resource management. It also has synergy with your cloud policies for defining risk and having a process to manage this risk. This will be covered more in *Chapter 3, Azure DevOps*, but it should be a consideration at the planning stage.

The key application architectures

Without further ado, let's work through some of the main architectural approaches for different application ecosystems. For each of these architectures, we'll also discuss the five principles you should support in your cloud governance:

- The cost management of the solution
- How to define a security baseline
- How to define resource consistency
- How to define an identity baseline
- How to accelerate deployment of the solution

You can also find a wealth of architectural information on Microsoft's Azure documentation pages: `https://bit.ly/35Dnzo7`.

Architecting a microservices ecosystem

Microservices have become a popular architectural style in application ecosystems, as they help provide a decomposed, highly scalable, resilient, and simple deployment model that can evolve quickly when necessary. Today's world of smaller functional blocks rather than large monolithic application tiers has found a friend in the cloud. This friendship is based on a resource's ability to be independent and scale on its own. The agility to quickly deploy and test is another big advantage provided by moving them to the cloud. The biggest problem is that these smaller autonomous pieces can create an operational nightmare for organizations. Considering that, let's look at how we approach these types of ecosystems in the cloud.

One benefit of the microservices strategy is that it enables services to be developed and deployed independently, meaning that faults can be isolated to one service rather than the application as a whole. These services aren't limited to a technology stack and they offer the ability to scale them independently from the application. This does, however, create a significant amount of complexity for development and testing, while also having a lack of overall governance and service latency. With these services depending on other services within the ecosystem, management, and versioning with change control can be cumbersome even for the most seasoned of organizations.

Figure 5 shows how an ecosystem like this could work:

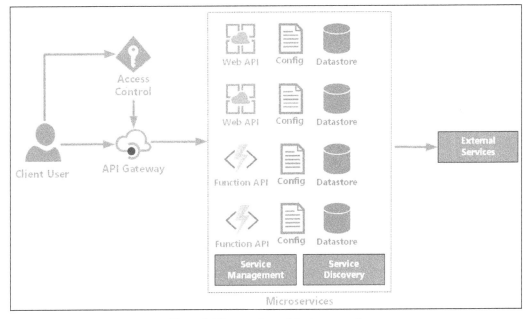

Figure 5: Microservice architecture sample

As you can see, the characteristics of a microservice architecture are:

- Each microservice contains code, configuration, and a data repository in the scope of the service, which provides a smaller footprint

- Each component in this hierarchical application model can be scaled, versioned, and upgraded independently

- These services can be written with different code and managed by different teams

- Services are split up based on business capabilities, making them smaller and with singular responsibility

The management and discovery services are a big part of the architecture; the management services are in charge of balancing and failure mitigation, while the discovery services help with discoverability. We use the API or Front Door to isolate clients from microservices, allowing better versioning, SSL termination, logging, centralized authentication, and load balancing.

Before we discuss the benefits and best practices, let's discuss the five principles that we mentioned at the top of this section:

- The cost management around microservices will depend on the direction you pick to host your services. If you use a serverless model, you will have the greatest cost-saving potential because of their consumption pricing. If you use a web app as your deployment strategy or require an app service-plan-backed function, your price structure will be reasonable and more predictable but costs will increase. If you choose to use Docker or a container approach, you will have an uptick in consumption pricing. You can choose to use **Azure Container Instances (ACI)**, **Azure Kubernetes Service (AKS)**, or Service Fabric, which may also require some extra management components and security configurations that you should plan on.

- The security baseline is based around the security needed to access the resource that you chose above, as well as the access to the endpoint exposed for the microservice, that is, HTTP/HTTPS. As we always recommend, if your service handles *any* sensitive information, use HTTPS to secure the endpoint.

- The resource consistency is based on the strategy chosen above. You should not choose multiple ways to deploy and manage your resource, so pick a predictable approach that can provide guardrails for your organization that cover all the development, deployment, and management risks that have been identified.

- The identity baseline will be how someone gains access to the microservice endpoint, like using API management to control and manage access. You can use shared keys or AD authentication to manage access to the microservice endpoint. Whichever you choose, you should work on persisting this approach to strategy as a whole. It is fair to point out that while shared keys are easy to use, they aren't individually tracked very well, and, when compromised, the keys in question must be rotated and all applications or services using them must also be rotated. I would always recommend using Azure AD over shared keys for this reason, and because of the added tracing and tracking you can do at an individual level for misbehaving actors.

- The deployment acceleration should adhere to an Azure DevOps flow and allow easy updating. Leveraging a solution with slots like functions or Web Apps or managed updating services like AKS or Service Fabric are extremely helpful in minimizing service downtime, if not eliminating it.

There are quite a few benefits to choosing a microservice architecture, as you can see below; however, it can complicate your support model if you do not plan out your guidance.

We can sum up the benefits of a microservice architecture as follows:

- Small and independent services
- Loosely coupled
- Separate code base
- Smaller deployments
- Internal dependency boundaries

Some simple best practices when operating under this architecture include:

- Decentralize everything
- Use the best technology for each service
- Define your API communication well
- Do not couple services
- Minimize cross-cutting concerns
- Isolate service failures
- Keep services from being chatty

While the microservice architecture was designed around business capabilities, one of the biggest challenges you have to overcome is defining the service boundaries. The general rule of thumb is one business action per microservice that is self-contained, which will require a significant amount of forethought. You will need to nail down your requirements, business domain, and overall goals in order to produce the right design. Leveraging something like domain-driven design can help with the strategic and tactical design, with the strategic design providing the structure of the system and the tactical design providing the entities, aggregates, and domain services of the domain model.

If you would like to read further on this topic, please refer to the online documentation: `https://bit.ly/37WqZDR`.

Architecting an event-driven environment

Event-driven architecture is not a new concept in Azure, and it has been around for many years in the on-premise environment within server-based applications like BizTalk, where it has been commonly referred to as "message-first development." As you can see in *Figure 6*, the event has a simple construction. It consists of a producer, ingestion or pump, and consumer, which is basically translated to this: an event is generated, there is a consumer listening for that event, and there is some glue in the middle.

Think of it like a phone: the person calling is the producer, you answering are the consumer, and the phone is the broker (ingestion). It is important to know that the producer and consumer are decoupled from each other in this simple model:

Figure 6: The event-driven architecture template

There are two types of event-driven architecture:

- **Pub/sub**: An event is published and sent to a subscriber, broker-style, so the event isn't replayable.

- **Event streaming**: This is a message pump where events are written to a log and are strictly ordered and durable. The client can read any part of the stream and is responsible for managing its position in the stream, which means clients can join anytime and events are replayable.

Good use cases for event-driven architecture are:

- Where the same events are used by multiple services or subsystems

- If near real-time processing is required

- If there is a need for complex processing, like aggregation or pattern matching

- High data velocity

Before we discuss the benefits and best practices, let's discuss the five principles that we mentioned earlier in the chapter:

- The cost management for an event architecture is based on Event Grid and your understanding of attaching the streams of data that are created.

- The security baseline is based on the event publisher's requirements, which usually aren't extremely high.

- Resource consistency is easy as there are not too many options other than Event Grid, Service Bus, Event Hub, and IoT Hub. These resources receive a message and it is the consumer's responsibility to receive and handle the message.

- Identity baseline is not overly applicable.

- The deployment acceleration should adhere to an Azure DevOps flow and allow for easy updating.

Let's look at some benefits of using an event-driven architecture:

- Highly scalable/distributed systems
- Decoupling of producers and consumers
- Consumers have independent processing of events

And here are some challenges you might face using this type of architecture:

- Processing order, especially in multiple consumer instances, as the events are streamed and it is the consumer's responsibility to manage their place in the stream.
- Guaranteed delivery is not part of this architectural structure because the producer doesn't care about the downstream consumer's ability to receive or process the events.

More documentation for event-driven architecture can be found at `https://bit.ly/39UR0Fr`.

Architecting a serverless ecosystem

The term "serverless" might be a little misleading, as you might think that these pieces of code magically run without a server. Contrary to its name, serverless doesn't mean these are magic pieces of code, but rather that the required resources have been abstracted from your view, meaning that you only have to worry about the code, logic, and integration of your application, and not the infrastructure that runs it, as you can see in *Figure 7*:

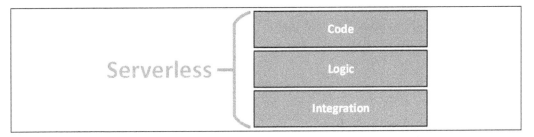

Figure 7: The serverless architecture template

Basically, serverless enables you to build cloud applications with minimal code, without the need to build, host, monitor, or maintain complicated resources or infrastructure.

There are currently three types of serverless resources in Azure:

- Functions, which are small bits of code to run in response to a variety of events
- Logic Apps, which are process steps for how to run your process (via workflows)
- Azure Event Grid, which is used to run your serverless infrastructure, as it helps stop polling on resources with an event-based architecture

Good use cases for a serverless architecture are:

- Where you want to use a microservice approach to deliver your service layer. This would follow the microservice architecture discussed earlier
- Needing to run small time-based processes
- Where flexible scalability is required, usually when you're using either a consumption model or an App Service plan
- Where you require utility-based billing; pay for what you use

Let's examine serverless architecture with regard to the five principles:

- The cost management for serverless is unique as they support a utility pricing consumption model and more predictable pricing with App Service Plan backing for functions. Most serverless solutions use an execution pricing for consumption, like 1m executions for a small price for Logic Apps and Azure Functions. For Logic Apps, you also need to include your connector cost based on the connectors you use.
- The security baseline is based on the security needed to access the resource that you chose above, as well as the access to the endpoint exposed for the microservice, that is, HTTP/HTTPS. Recommendation: if your service handles ANY sensitive information, use HTTPS to secure the endpoint.
- The resource consistency is based on the strategy chosen above. You should not choose multiple ways to deploy and manage your resource, so pick a predictable approach that can provide guardrails for your organization that cover all the development, deployment, and management risks that have been identified.
- The identity baseline will be how someone gains access to the microservice endpoint, like using API management to control and manage access. You can use shared keys or AD authentication to manage access to the microservice endpoint. Whichever you choose, you should work on persisting this approach to strategy as a whole.

- The deployment acceleration should adhere to an Azure DevOps flow and allow for easy updating. Leveraging a solution with slots like functions or Web Apps or managed updating services like AKS or Service Fabric are extremely helpful in minimizing service downtime, if not eliminating it.

Now let's look at some of the benefits and challenges that come with serverless. One of the biggest challenges that you'll come across is that serverless solutions are usually vendor-specific and can require code rewriting before moving them between on-premise or other clouds.

The benefits of using serverless architecture include:

- Independent cloud-based logical functions
- Stateless by design
- Event-triggered
- Fully managed by the cloud provider
- No system administration

Challenges you will face using serverless architecture include:

- Reduction in overall control other than the specific code needed to run the serverless applications
- They are vendor-specific in both code structure and underlying resources
- The cost can be unpredictable as you pay for what you use
- You need to have discipline to help fight against service or code sprawl
- They have an execution cap and should be designed for small, quick-running processes

With the introduction of serverless, a micro-billing or utility-based billing structure was also introduced. This was a unique deviation from the current subscription-based model in Azure and was built around consumption.

For Azure Functions, there are three types of billing:

- Consumption or pay-as-you-go
- Premium plan
- App Service plan

For Logic Apps, there are two types of billing:

- Actions
- Connections

For Event Grid, there is one type of billing:

- Number of events processed

Here's how the serverless architecture relates to our five principles:

- The cost management for serverless is unique as they support a utility pricing consumption model. Almost all serverless use an execution pricing for consumption, like 1m executions for a small price for Logic Apps and Azure Functions. For Logic Apps, you also need to include your connector cost based on the connectors you use.

- The security baseline is based around the security needed to access the resource that you chose above, as well as the access to the endpoint exposed for the microservice, that is, HTTP/HTTPS. Recommendation: if your service handles ANY sensitive information, use HTTPS to secure the endpoint.

- The resource consistency is based on the strategy chosen above. You should not choose multiple ways to deploy and manage your resource, so pick a predictable approach that can provide guardrails for your organization that cover all the development, deployment, and management risks that have been identified.

- The identity baseline will be how someone gains access to the microservice endpoint, like using API management to control and manage access. You can use shared keys or AD authentication to manage access to the microservice endpoint. Whichever you choose, you should work on persisting this approach to the strategy as a whole.

- The deployment acceleration should adhere to an Azure DevOps flow and allow for easy updating. Leveraging a solution with slots like functions or Web Apps or managed updating services like AKS or Service Fabric are extremely helpful in minimizing service downtime, if not eliminating it.

Let's take a moment to review some of the features, best practices, and considerations of the three Azure serverless resources.

Azure Serverless Functions

Here are some best practices for using Azure Functions:

- Avoid large long-running functions – Refactor large functions into smaller function sets that work together and return fast responses, unless using durable functions.

- Cross-function communication – When integrating multiple functions, it is generally best practice to use storage queues for cross-function communication.

- Write functions to be stateless – You should associate any required state information with your data.

- Write defensive functions – You should assume your function could encounter an exception at any time, so design your functions with the ability to continue from a previous fail point during the next execution.

- Don't mix test and production code in the same function app.

- Use asynchronous code, but you should avoid call blocking by not referencing the `Result` property or the `wait` method on the instance of the task.

Logic Apps

Things you should consider before using Logic Apps:

- Use the message-first development pattern. Some patterns can be found at `https://bit.ly/2t2WPjH`.

- Logic Apps use the **If This Then That** (**IFTTT**) architectural pattern, which means you use conditional statements and event triggers.

Logic Apps provides a variety of triggers and actions with management tools to help centralize your API development. They consist of:

- **Workflows**: A graphical way to model your business processes as a series of steps or a workflow.

- **Managed connectors**: These connectors are created specifically to aid you when you are connecting to and working with your data.

- **Triggers**: Some managed connectors can also act as a trigger. A trigger starts a new instance of a workflow based on a specific event, like the arrival of an email or a change in your Azure Storage account.

- **Actions**: Each step after the trigger in a workflow is called an action. Each action typically maps to an operation on your managed connector or custom API apps.

- **Enterprise Integration Pack**: For more advanced integration scenarios, Logic Apps includes capabilities from BizTalk, Microsoft's integration platform. The Enterprise Integration Pack connectors enable you to easily include validation, transformation, and more into your Logic Apps workflows.

Event Grid

Event Grid is an Azure service for managing event routing within your system. It is particularly useful in serverless architectures, where it can do the hard work of connecting your data sources to your event handlers. It's well worth understanding some of the concepts and terms underpinning Event Grid so you can assess it for your needs:

- "What Happens" is the event. An example is if Event Grid is attached to a subscription and a security role is added; an event would then be raised to Event Grid.

- "Where did this take place" is the event source. This could be a subscription, for example. It is the object that has the event.

- "Publisher Endpoint" is referred to as the topic. This is the endpoint that the event is published to.

- "Filtering of incoming events or event routing" is the event subscription, the consumer of the event.

- "Thing that reacts to an event" is the event handler. This is the consumer action to the event.

- Security is handled through auth keys or SAS tokens.

- Retry is built in for confirming event receipt.

- For batching publishing events together, batches up to 1 MB can be accepted, with each event recommended to be less than 64 KB.

Now that we've covered the basics of a serverless system architecture, let's look at a mobile application architecture on Azure.

Architecting mobile applications

While mobile development is not cloud-related per se, it makes sense to add a quick description of what developing a mobile solution would look like. When developing a mobile application, you need to consider compatibility with a variety of platforms, so using a development platform like Visual Studio with Xamarin is recommended. It allows you to create native experiences with a single code base. Note that you will still need Apple hardware, like a MacBook, to compile and upload to the Apple App Store, but a single development environment really simplifies the development process.

This will give you the opportunity to leverage all your favorite cloud services in an environment you are used to, as you can see in *Figure 8*:

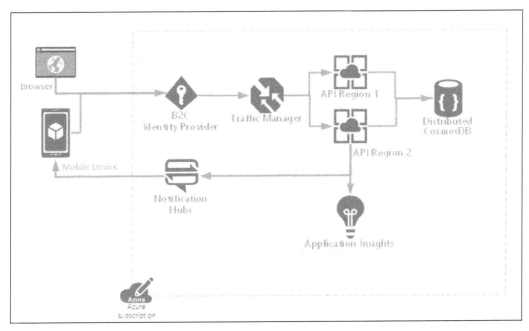

Figure 8: Simple mobile architecture

Let's consider mobile architecture using the five principles we covered earlier:

- The cost management for mobile application development encompasses your security model, **Azure Active Directory (AAD)**, B2C, and so on, and your notification services. You will need to factor in back-end services and data repositories.

- The security baseline is based around the security needed to access the resource that you chose above, as well as the access to the endpoint exposed for the microservice, that is, HTTP/HTTPS. Recommendation: if your service handles ANY sensitive information, use HTTPS to secure the endpoint.

- The resource consistency is based on the strategy chosen above. You should not choose multiple ways to deploy and manage your resource, so pick a predictable approach that can provide guardrails for your organization that cover all the development, deployment, and management risks that have been identified.

- The identity baseline will be how someone gains access to AAD or B2C from a device and using API management to control and manage access. Whichever you choose, you should work on persisting this approach to strategy as a whole.

- The deployment acceleration should adhere to an Azure DevOps flow and allow for easy updating. Leveraging a solution with slots like functions or Web Apps or managed updating services are helpful in minimizing service downtime, if not eliminating it.

While mobile solutions aren't really considered cloud-related, but are instead device native, they rely on the cloud for their back-end processing services and management. It is nice to see the mobile development and back-end service development environment aligned within the Visual Studio ecosystem. Online documentation for Xamarin can be found at `https://bit.ly/2FOlIm8`. Let's now turn to another common cloud architecture use case, IoT.

Architecting an IoT ecosystem

IoT applications are normally described as the things or devices that send data to generate insights that fire actions to perform a process. A good example is using a sensor to tell if a freezer door is open. When the sensor detects the door is open, it sends insights that generate an action of notifying someone that the door is open.

IoT Central, at `https://docs.microsoft.com/en-us/azure/iot-central/`, is a managed SaaS-based solution for managing your IoT devices. It doesn't enable as much customization as a PaaS-based solution, but it is easy to get started with. IoT devices are all about telemetry, basically reading sensors to measure and get information about the IoT device itself, as you can see in *Figure 9*. There are two ways to process telemetry as it is being sent:

- Short-term storage, or "hot path," is near real-time telemetry and is usually handled by a stream process engine, which usually creates the action of alerting or logging for analytic tools to query.

- Long-term storage, or "cold path," uses interval-based processing and usually deals with larger volumes of data, often to be processed by machine learning to find further defined business process actions.

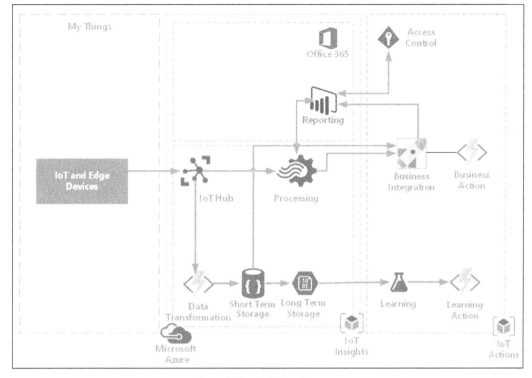

Figure 9: Simple IoT architecture

Let's look at the list of things we have used in our simple architecture:

- IoT devices, which are the things we securely connect to the cloud
- IoT Hub or gateway is the secure connection for data ingestion
- Stream processing is when the data is analyzed as it comes through the gateway
- Short-term storage or warm path is a short-term holding pen for incoming data
- Long-term storage or cold path is a long-term holding pen for historical data
- Data transformations help provide a consistent data shape for storage and processing engines
- Business processing is the action taken on the insights
- User management is the access control actions performed in Azure and on the IoT devices, like upgrades

These are some things you need to understand that affect scaling an IoT architecture:

- Remember your daily quotas – https://bit.ly/2FPYZ8Y
- Understand the device quota of the connected devices
- Keep a close eye on ingestion throughput and process throughput

Use parallelization in stream analytics to split the workload across nodes. There is a maximum number of function instances per IoT Hub partition, so it is recommended to process the message in batches.

There are some specific security considerations that you should keep in mind with IoT devices:

- Implement a secure data encryption policy
- Provide a digitally signed data encryption policy
- Ensure it supports TLS 1.2 and DTLS 1.2
- Ensure key-store and device keys are updatable
- Ensure device updating, like firmware and app software, is part of the process

To help with troubleshooting, you want to enable monitoring and logging as much as you can with the subsystems. Monitoring and logging can provide answers to the following operational questions:

- Is there an error condition?
- Is something misconfigured?
- Is data accurate?

Monitoring systems help provide insights into the health, security, stability, and performance of an IoT solution. Some of the metrics to collect might be:

- IoT devices and subsystems reporting configuration changes
- Data repository read and write performance, schema changes, security audit logs, locks or deadlocks, index performance, CPU, memory, and disk usage
- Managed services health reporting and configuration changes that impact dependent systems

You can find out more about IoT architectures with Azure at https://bit.ly/2Tb5Ulg.

Let's now turn to our final architecture for examination, web-based applications.

Architecting web-based applications

You should always start web-based solutions with maintainability in mind. You should also start from a touch-first perspective, as touch-based devices are becoming more popular, which should lead to a more responsive design overall. So, let's start by reviewing a list of the characteristics of modern web application design:

- Cross-platform capable
- Cloud-hosted
- Scalable (autoscalable preferred)
- Modular in design (smaller pieces)
- Loosely coupled
- Easily testable (automation preferable)
- Responsive design
- Easy to deploy

Now let's review some common design principles that will help achieve those characteristics:

- **Encapsulation**: Limiting the amount of outside access to an object's internal state. So, if an actor wants to change the state of an object, it does so through a well-defined function or property setter, not through direct access to the private state of that object.

- **Separation of concerns**: The basic principle that applications should be separated based on the work performed. This usually boils down to separating core business behavior from user interface logic – simply stated, don't put business logic in your UI.

- **Dependency inversion**: The OOP principle that high-level modules should not depend on low-level modules; they should both depend on abstractions. Focus on the abstractions – the interfaces – first, rather than the implementation details.

- **Persistence ignorance**: Classes should be designed purely to solve the business problem of the moment, and should not be diluted by concerns with persistence.

- **Bounded context**: A pattern from domain-driven-design where models are separated into different domain contexts with a strict relationship. This enables different contexts to evolve differently and have different lexicons while maintaining a relationship.

Remember as you design your web application to keep simple application layering in mind to help with the separation of concerns, as you can see in *Figure 10*:

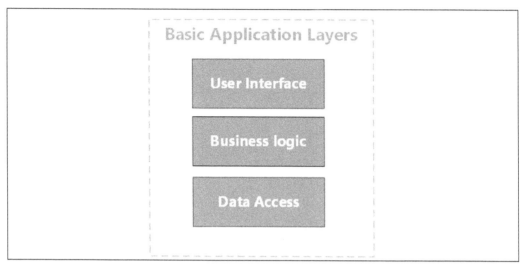

Figure 10: Basic application layers

Here are some web application best practices:

- Co-location should be in the same region to help reduce latency and monetary charge for cross-region transfers.
- Enable the App Service Auto-Healing feature to help your web application remain healthy.
- Consider autoscaling to help combat CPU pressure.
- Load test your web application.
- Use the deployment slot when you can.
- Enable diagnostic logging.
- Enable monitoring.
- Decouple your app logic.
- Automate your infrastructure with ARM templates.

Online documentation for web application architectures can be found at `https://bit.ly/2R5VRLk`.

Now that we've covered the main types of application architecture you'll use with Azure, let's turn our minds to some design best practices that apply to all of the architectures.

Architecture design best practices

In order to start our discussion on best practices, we need to start with security. As we saw in our responsibility matrix back in *Figure 2*, regardless of the type of service deployed you will always keep security responsibilities over:

- Accounts
- Access
- Endpoints
- Data

Security is a critical element of your journey to the cloud. Understanding your data and classifying it into categories based on sensitivity and business impact will help provide insight into optimizations that your business may not have realized when data was unclassified. While data classification is not a silver bullet that will resolve all of your problems, it can surface benefits like compliance improvements, and improve ways to manage these resources. This also helps your organization mitigate risk through rights management, encryption, and data loss prevention, which can come at a large cost, based on retention policies, if your organization does not plan for it. These types of policies can be implemented in Azure through Blueprints and Compliance Manager, which can help organizations set up a repeatable, standardized approach. Azure Blueprints provides a way through management groups to help deploy and govern the subscriptions and the resources within those subscriptions, while Compliance Manager provides a more holistic view of your compliance posture and overall data protection. Management groups also allow policy management at all levels within your management group structure. These are key concepts that will act as the foundation for your journey to the cloud.

Let's quickly dive into Blueprints and management groups in a little more detail to provide a basis for why and when you would use them. Management groups provide a centralized approach to managing security, deployments, and policies that exist in a level above subscriptions.

Note that management groups start at the tenant, which is the root, and can contain a limited number of trees or nested groups. These limits and structures can be reviewed at `https://bit.ly/2sa6MLI`. It is worth pointing out that your structure for management groups should be created so that they remove the need to manage RBAC at the resource or resource groups level.

Let's look at what a simple management group structure might look like:

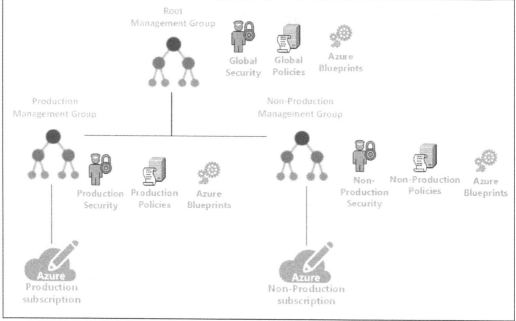

Figure 11: A Simple management group structure

As you can see in *Figure 11*, there is a root management group and two branches separating production from non-production. If you have a security group, policy, or blueprint that spans branches, then put it in the highest branch possible. For example, if I have an architecture group that requires owner access to all branches, then I would add them to the root tenant. This is because management groups inherit in a parent/child model.

As you can see, management groups allow you to define your security, policies, and blueprints (deployments) at any level. For instance, I can add the architecture group at the root level because their access spans all subscriptions, but I might only add the developer group to the non-production branch. This also transitions to policies, as I might want to limit the size of VMs in the non-production branch to help control cost. Blueprints in management groups could be applied to help deploy environments and cost by assigning the development version to the non-production branch and a production version to the production branch. Now that we've covered the concepts of why you might want to use management groups in your cloud security, let's list some general security best practices:

- Use management groups to manage security, initial deployments, and policies

- Use Security Center Standard, at least, for your production subscriptions
- Use Azure Key Vault to store secrets and keys
- Use **Web Application Firewall (WAF)** to help manage exploits and vulnerabilities
- Use **Multifactor Authentication (MFA)** to protect organization authentication methods
- Use encryption on all storage resources, SQL, files, Blob, VM disk files, and so on
- Use VNets to isolate VMs, appliances, and other devices

With those basic security best practices in mind, let's now run through some more specific best practices for databases, Platform as a Service (PaaS), and Infrastructure as a Service (IaaS) in Azure.

Here are some database security best practices:

- Use firewall IP rules to limit access, and turn off the flag allowing all Azure resources access
- Use authentication mechanisms like AAD to prove a user's identity; limit using SQL authentication as much as possible
- Leverage an authorization mechanism to limit access, and use the principle of least privilege
- Use **Transparent Data Encryption (TDE)** to encrypt your database files
- Enable auditing and threat detection for sensitive data

Online documentation for database security can be found at
`https://bit.ly/2NbGrnW`.

Here are some PaaS best practices:

- Use Key Vault for secrets
- Use Application Settings for app settings whenever possible
- Protect and monitor all endpoints, internal or external
- Secure all endpoints and applications using strong authentication and authorization
- Always use defensive coding and build for fault tolerance
- Use IP restrictions or WAF to secure endpoints

Online documentation for PaaS security can be found at `https://bit.ly/37QX53X`.

IaaS best practices:

- Control your VM access and apply policies for compliance
- Use DevOps ARM templates to simplify setup and deployment for repeatability
- Use a least-privilege approach for privileged access
- Use Security Center and install antimalware
- Use availability sets for high reliability
- Always keeps VMs current
- Always encrypt VM disks
- Monitor VM for threats and performance

Online documentation for IaaS security can be found at `https://bit.ly/2T9T1b6`.

As you can see, most of the best practices relate to security. This is because with most cloud resources quite a few things are abstracted from your view, creating an opportunity to focus more on your security solutions rather than being distracted by the underlying sources. The exception to this is IaaS, where you still have quite a bit left in your view.

That wraps up our run-through of application architecture types in Azure. You can review the full suite of documentation at `https://bit.ly/2QFJF1E`. Next, let's look at scalability and application management in Azure.

Design principles for scalable and manageable applications on Azure

Applications in the cloud have to be able to respond to issues or faults such as unavailability, data or network loss, time-outs, or service transition. Some of these issues can be temporary and a basic retry may overcome them, while others will take more work. The first thing to do is to create your applications with resiliency and self-healing in mind.

The foundations of a self-healing system consist of:

- Automatic detection of the issue

- Taking action to respond to the issue detected
- Auditing all relevant information about the issue

Self-healing applications rely on designing your applications for resiliency, which means you need to plan for failure with minimal downtime and data loss.

Designing for resiliency

There are two main characteristics of resilient applications:

- They can recover gracefully from failures and experience minimal downtime
- They run in a healthy state with no significant downtime

The biggest difference with resiliency in the cloud is that you can scale up as well as out.

Here are a few ways to help you build for resilience:

- Define your requirements based on workload decomposition and business needs:
    - Document usage of your workloads
    - Plan for usage to ensure uptime
    - Define an availability and recovery matrix for your SLAs

- Use architectural best practices to meet your business requirements:
    - Perform **Failure Mode Analysis (FMA)** to help identify the types of failures you might experience
    - Create a redundancy plan for each workload, including costs
    - Plan for scale, keep limits in mind
    - Plan a subscription and resource map
    - Plan a management group structure
    - Manage data over all storage, backups, and replications

- Test, test, test, and perform forced failovers:
    - Test common failure scenarios
    - Perform load tests
    - Run application fire drills
    - Test health probes

- ○ Test monitoring
- ○ Test external services and how they are monitored

- Deploy consistently and leverage automation early:
  - ○ Automate application and infrastructure deployment early in your process
  - ○ Log and audit deployments
  - ○ Document processes and releases

- Monitor health to detect failures and send alerts:
  - ○ Implement check functions and health monitors
  - ○ Maintain application logs
  - ○ Check long-running workflows
  - ○ Track transient exceptions and retries

- Have a recovery plan:
  - ○ Plan on support interaction with the cloud provider
  - ○ Have a disaster recovery plan
  - ○ Prepare for application failures
  - ○ Have plans to:
    a. Recover from data corruption
    b. Recover from service failure
    c. Recover from region-wide failure
    d. Recover from a network outage

Architectural overview and considerations

As we have discussed, security is at the foundation of cloud principles, and key to this in Azure is understanding how RBAC works. As we showed earlier in the chapter, leveraging management groups is also extremely helpful in your subscription access management, and with RBAC you only provide users the minimum amount of access needed to complete their jobs. This can also be coupled with **Privileged Identity Management (PIM)** should the need arise for a member to elevate their access or "just-in-time" their access within a given environment for a given time, which helps with approvals and auditing.

It's worth getting more details on the following list of tools and resources that are available in Azure to aid in security:

- RBAC
- Antimalware
- MFA
- PIM
- ExpressRoute
- VPN (alternatively called Virtual Network Gateway)
- Identity Protection
- Security Center
- Intelligent Security Graph

Identity management and Azure AD

AAD is the Azure solution for identity and access management that backs a tenant. AAD was designed as a multi-tenant solution that is cloud-based and provides identity management services. It combines core directory services, application access management, and identity protection into a single solution.

Some best practice recommendations for identity management in AAD and access control security are:

- Treat identity as the primary perimeter for security
- Centralize identity management by having a single AAD instance that is synchronized with on-premise directories, and make sure to enable password hash synchronization when you do
- Leverage AAD in all new development projects requiring identities
- Manage connected tenants by assessing the risk
- Enable SSO for users
- Turn on PIM for elevation privileges
- Enable password management for your users
- Enforce MFA for users to help mitigate identity theft
- Use RBAC in conjunction with management groups

Let's look at how a simple hybrid model might work in *Figure 12*. As you can see, we have an on-premise Active Directory (AD) and an AAD. They use AAD Connect to synchronize identities so that users can leverage SSO to access both the corporate network and the cloud-based application:

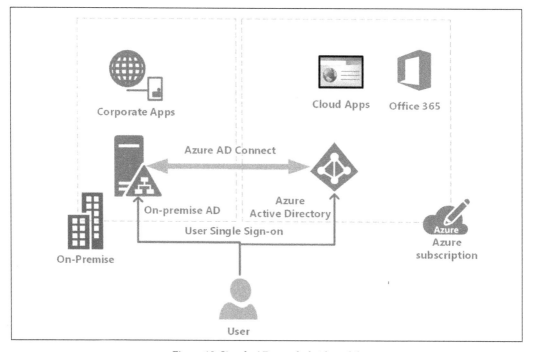

Figure 12: Simple AD sync hybrid model

When to use B2C and B2B

AAD **Business to Customer (B2C)** and **Business to Business (B2B)** were created to work with external identities within AAD. B2C handles customer-facing applications with self-serve and social logins, and B2B is for business-facing access. These function almost like **organizational units (OUs)** within an on-premise AD, and they separate these types of users from your main corporate AD. Here is a simple way to decide when you should choose to use them:

- When you need to authenticate users to an external organization and leverage work or school accounts, use B2B

- When you need to invite customers to your web or mobile app and you would like to also allow social logins, use B2C

Data protection

We should start by understanding what is provided by default from the cloud provider and the access-control requirements that are established by the Azure security policy, which states:

- No access to customer data, by default

- No user or administrator accounts on customer virtual machines

- Grant the least privileged access to complete the task and audit

Data protection is provided through data segregation, which logically isolates customers from one another. TDE provides further security by encrypting the data at rest, which spans SQL databases to VMs. Data in transit is also encrypted from the customer to the cloud as well as between cloud systems. You also have the ability to further protect your data through data redundancy for both in and out of country, and, at a smaller scale, in and out of regions, depending on your compliance. This means that data will be replicated with the defined compliance ranges and three types of redundancy are offered:

- **Locally redundant storage (LRS)**: This option makes three copies of your data in a single facility in a single region

- **Zone-redundant storage (ZRS)**: This option makes three copies of your data across two or three facilities with a single region or across two regions, which provides durability within a single region

- **Geo-redundant storage (GRS)**: This option makes six copies of your data with three copies within the primary region and three copies within a secondary or sister region, which provides durability within two separate regions

Data destruction in the cloud has strict policies of ensuring overwriting of storage resources before they are reused. Customer data is never inspected, approved, monitored, nor claimed as being owned by a cloud provider; this is the responsibility of the customers. This also means that the customer owns data retention.

Networking

Networking in the cloud doesn't differ too much from on-premise or datacenter networking except for in isolation, availability, and scaling. You have to remember that address spaces are made up of subnets that are contained within subscriptions. VNets are good for isolating resources in Azure as well as giving a landing spot for VPNs or ExpressRoute to on-premise or other cloud provider networks.

Let's take a moment to review some VNet concepts:

- IP address spaces are large buckets of acceptable IP addresses
- Subnets are portions of the address space of IPs that segment the IPs into smaller chunks that can be controlled by different **Network Security Groups (NSGs)**
- VNets can only live in a single region and are scoped to the subscription, which is a hard boundary

Here are some best practices for using VNets:

- Try not to overlap your address spaces.
- Subnets should not span the entire address space.
- Keep your VNets to a minimum.
- Use NSGs to secure your subnet in your VNets. You can use **Application Security Groups (ASGs)** within NSGs to help further control the network flow to resources within the subnet.

When communicating with resources in the cloud, you can connect resources directly to VNets, creating an isolated boundary. You can also use a service endpoint in VNets to secure resources without needing a direct connection. Connecting VNets together can be done via peering. Connecting to the cloud from on-premise to VNets is also an option. You can connect from a VNet to a single resource on a corporate network through a point-to-site VPN connection. You can also establish a full site-to-site VPN for connecting to a corporate network or use an ExpressRoute, which is a circuit between the cloud and your network. *Figure 13* shows a hybrid network that is using an ExpressRoute that requires an underlying third-party connection to the cloud. The network is also using a site-to-site VPN as a backup should the main circuit go down:

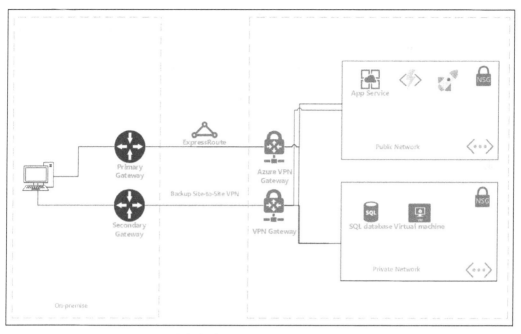

Figure 13: A simple hybrid network between on-premises and the cloud

You can leverage NSGs that will filter traffic for inbound and outbound security rules. This can also be done with appliances such as firewalls. In our example, we use a simple NSG to block private VNet VMs from accessing the internet.

Azure for containerized apps

The first thing that you always run into when discussing containers is, why should I care about containers? This is usually answered in one of two ways. Firstly, containers provide the freedom to move your application from on-premise to the cloud or within the cloud to another cloud provider with no code change to your application. Secondly, each application is self-contained, meaning all application elements and their versions are contained within the boundaries of the container, so changing a library for an application won't negatively affect or spark a redeployment of all the applications that shared the library.

So, what are containers? I like to use a shoebox analogy to describe containers. A shoebox is pretty standardized and small. You can only fit so much in them, and they tend to have a single function like storing your photos. You can store this shoebox and move it around pretty easily. It does, however, require a shelf or floor to stack on. Containers, like these shoeboxes, are a standardized package of software that contains everything it needs from code, runtime, tools, libraries, and settings. They can also be moved around easily, but like shoeboxes they require some infrastructure to stack upon. This might sound a bit like our discussion of microservices; however, containers can house a bit more. *Figure 14* is an example of a simple containerized application:

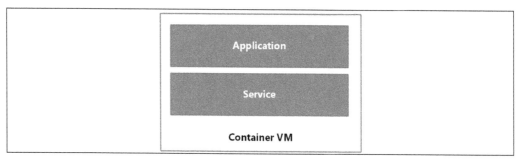

Figure 14: A simple container application

One of the most standardized containerization technologies used in the cloud is Docker, which has been around since 2008. It is open source and enables you to package, ship, and run your applications in all current cloud environments. Now, before you jump headlong into using containers there are a few key points to discuss. Remember we said that you need some infrastructure? This is referred to as orchestration or the orchestrator, and this is where Kubernetes in the cloud comes into play. As a side note, I would only use container instances in a proof of concept (POC) or a quick application as they do not contain an orchestrator. This technology can scale your containers up and down based on usage. The orchestrator can roll out changes, and it can roll back instances that fail or don't pass health checks. It also helps simplify management and load balancing to containers or a Docker swarm.

Azure container tools and services

Azure provides several platforms for supporting containers:

- AKS
- Azure Red Hat OpenShift
- ACI
- Web App for Containers

Visual Studio and Visual Studio Code have also added support for containerizing your .NET Core code. Docker has also provided support for Windows containers to further the .NET ecosystem. Once the workload is active, you will have the ability to containerize your .NET Core code by simply right-clicking and selecting **Add Docker Container**.

Azure Red Hat OpenShift

This provides a single pane of glass that extends Kubernetes by providing additional tools and resources to help with images, storage, networking, logging, and monitoring. In other words, it gives you the ability to choose your own storage, image registry, networking, and other tools to help with automation. It also helps provide better integration with AAD and Kubernetes RBAC in conjunction with monitoring the health of your resources and clusters. You can get more information at https://docs.microsoft.com/en-us/azure/openshift/.

Azure Container Instances (ACI)

Container instances are a really simple way to create an isolated container for small applications or tasks that do not require full orchestration. These instances have a fast startup time and allow for public IPs and DNS naming. Containers can be grouped and they leverage VNets. You should review and understand the limits found at https://bit.ly/30hsme5.

Container instances are the simplest way to begin working with containers, plus it is a great way to proof-of-concept applications from an architectural perspective.

You can find out more about ACI at https://bit.ly/2Tabwfj.

Azure Kubernetes Service (AKS)

AKS is a simple managed Kubernetes cluster in Azure. It removes the need to know how to set up Kubernetes in order to build out a container infrastructure. You will still also benefit from Kubernetes capabilities like health monitoring and maintenance. The Kubernetes masters are managed by Azure and are free; you only pay for the agent nodes within your clusters. You can also integrate AKS with AAD, allowing the use of Kubernetes RBAC. The monitoring data is stored in the Azure Log Analytics workspace, which can monitor cluster performance and the health status of workloads within AKS.

These are the key components to understand:

- Access, security, and monitoring – AKS uses RBAC to help control resources and namespaces. This can be connected to AAD. Access can be configured based on existing identity or group membership.

- Cluster and nodes – AKS nodes are run on Azure VMs and clusters run multiple node pools, so mixed operating systems are supported. You can use both pod and cluster autoscalers, which can scale automatically based on demand.

- AKS nodes support integration to VNets in most setups.

- Development tooling – There is a Kubernetes extension for Visual Studio Code as well as other tools like Helm and Draft.

- Docker support

- Ingress controller – This helps control traffic routing, reverse proxy, and TLS termination.

Online documentation for AKS can be found at `https://bit.ly/37TKSvv`.

It is also possible to leverage Docker containers for Web Apps. This is similar to a container instance. Online documentation can be found at `https://bit.ly/2QFR4BB`.

Summary

We've touched on a lot of topics in a short space of time, but hopefully you've gained an overview of the various architectures available for building applications with Azure. We've also looked at application design best practices, mostly focused on security, and finished with a look at working with containers in Azure. In the next chapter, we'll look at DevOps and how its methodology integrates with the cloud.

3

Azure DevOps

Introduction

DevOps is a way of helping provide a better software delivery experience. In this chapter, we will discuss what DevOps is and why you should use it in the cloud. You will see that there are quite a few benefits once you understand how to successfully implement it, and Azure has a lot to offer businesses taking a DevOps approach.

Why DevOps?

Many organizations making the journey to the cloud struggle to create a repeatable process for deploying their applications and change control. They struggle because of the lines of responsibility discussed in *Chapter 2*, *Architecture Choices and Design Principles*, and this is where DevOps and cloud adoptions have great synergy. DevOps can improve that deployment while reducing costs in infrastructure, code, monitoring, and policy deployments. Before we discuss how this works, let's first learn about DevOps and why it is important in this journey.

DevOps began as a software development methodology that accelerates the building, testing, and releasing of software applications by bringing together two main groups, the **developers** (**Dev**) and the **operations** (**Ops**), to work more effectively. It is not meant to be a replacement for any agile or lean methodologies, but more as a supplement to them. It fills in the gaps within tech companies to break down functional and non-functional specs, while also working to automate as much as possible in the spirit of speed and quality, and refine the operational process.

DevOps is a software development method that emphasizes communication, collaboration, integration, automation, and the measurement of cooperation between solution developers and IT professionals.

The DevOps process creates a cyclic process of planning, coding, building, testing, releasing, deploying, operating, and monitoring the applications you develop for your organization, bringing the development side and operations side together. This is the core representation of the DevOps culture shift, which has a more detailed process than the Microsoft Azure DevOps tool. You can see this cycle illustrated in *Figure 1*:

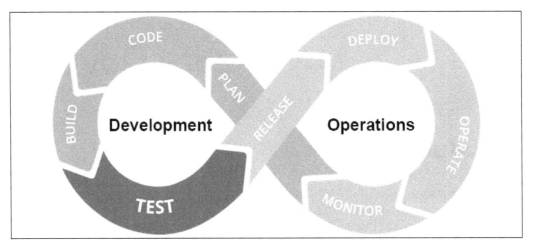

Figure 1: DevOps process loop

DevOps as a whole is less about what you do and more about how you do it. With evolving technologies and processes, DevOps principles are critical in your journey to transforming your organization. At the end of the day, however, it is about how work gets done and how people interact with each other and with the technology to drive performance. DevOps is not an off-the-shelf solution that can just be bought and bolted on, but Microsoft has released tools such as Azure DevOps that make integrating DevOps with Azure much easier. For Azure DevOps, this main paradigm around how DevOps was delivered to an organization changed, as you can see in *Figure 2* from Microsoft's Azure DevOps at `https://bit.ly/2FxOOHI`:

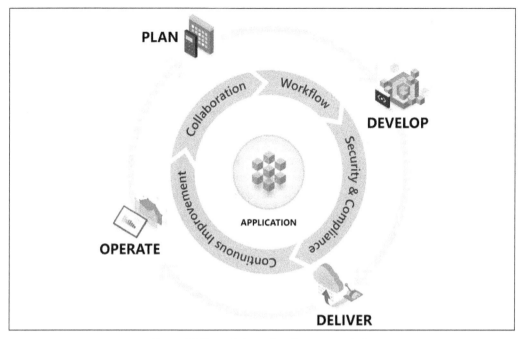

Figure 2: Microsoft Azure DevOps process circle

The main simplification is in the reflection on the tooling provided, which will be discussed later in this chapter. To quickly review how this simplification affects planning, development, delivery, and operations, the **PLAN** part helps the team visualize their workload while helping to provide a definition of the work and track the workload.

This helps to provide a smarter and faster development process that integrates with Visual Studio Code and Visual Studio, providing like-minded developers the ability to share and collaborate on their code, while automating testing, builds, and releases for a streamlined **continuous integration** (**CI**) process. This also leads to helping deliver your applications through the **continuous deployment** (**CD**) process, allowing environment-defined variables to provide a more standardized delivery process with significantly fewer deployment problems than manual deployments. Once these applications are delivered, you need to operationalize your environment with alerting, logs, and telemetry, which includes policies and compliance through Azure Blueprints within management groups.

Management groups are a way to standardize your subscriptions in Azure. You can learn more about management groups at `https://bit.ly/2T2wU6m`, which I would highly recommend. But that's enough of a detour. Let's get back to our DevOps discussion.

DevOps is about embracing communication by bringing together the operations side of the house with the developer side of the house, giving everyone a seat at the table throughout the software cycle. It is not an individual's job but rather everyone's job, and it's a critical way of working together to drive performance. The people component of the model seems to be the part that is least defined. It's important to realize that simply doing more stuff is not the solution to a successful DevOps transition, or any cultural transition for that matter, within technology.

The main purpose of DevOps is to allow developers and operations to overcome the challenges they face. Let's look at what this means. This is how developer and operations roles are traditionally seen:

Developers	Operations
• Feature delivered after manual unit testing on development systems • Rewarded mainly on timely delivery • Development systems not the same as production • Little to no concern about infrastructure/deployment impact related to code changes • Little to no feedback on changes until later in the software cycle	• Code delivered with little to no concern about infrastructure • Reward mainly for uptime • Left with solving security issues • Left with solving monitoring of the application with little to no knowledge of the application • Usually engaged at the tail end of the application development cycle

As you can see, the developers and operations are almost pitted against each other, mainly in how they are rewarded. This can cause an uphill struggle, with each group throwing grenades at the other in the form of tasks when their goals are conflicting. Neither side really understands the "why" behind the urgency of what is being asked. Some important attributes that need to help bridge this gap are:

- Release management, which provides a more in-depth understanding of risks, dependencies, and compliance issues

- Release and deployment coordination, better tracking of discrete activities, more consistent and faster escalation of issues, documented process control, and reporting

- Release and deployment automation, more consistent and repeatable processes that can be automatically invoked and executed with or without (non-production) an approval process

We'll be looking at how some of those things can be implemented in Azure later in the chapter. It's also important in DevOps to strongly define roles and responsibilities. Here are some roles within a traditional DevOps structure that facilitate the goals of accelerating company processes:

- **Evangelist**: The leader that works at implementing and orchestrating DevOps processes within the organization. They are responsible for keeping the process moving forward and hold the vision for the overall DevOps process and roles.

- **Release Manager**: Responsible for the project through production, where they oversee all the development, testing, and deployment to support continuous delivery. They provide visibility into the process by measuring and interpreting the metrics on all tasks.

- **Quality Assurance**: A key role in the successful delivery of the project that is responsible for ensuring the overall user experience and that the product is free from bugs.

- **Security and Compliance Engineers**: Ensure the project meets all the standards and regulations while ensuring the product is secured against attacks.

- **Developers**: Responsible for developing the production to meet the requirements of the business or client.

So, to concisely answer "Why DevOps?", we can define several benefits that are the most important to the technical, cultural, and business sides of the operation:

Technical Benefits	Cultural Benefits	Business Benefits
• Faster resolution of issues • Less complexity • Continuous delivery	• More productive teams • Higher project engagements • Greater professional development	• Fast delivery of features • More stable operating environments • Improved collaboration • More innovation time

Now that we have discussed what DevOps is, let's look at the methodology.

Azure DevOps – the methodology

Let's take some time to review a few things to remember during your DevOps journey. It's all about trust; development and operations teams are typically segregated from one another and unable to communicate, let alone collaborate, effectively, and DevOps aims to solve these issues. You must understand the people, which begins with understanding ourselves first and then those around us. This will lead to the end of the blame game. Expect to fail, fail often, and fail early, so embrace failure and learn. Work on clearing bottlenecks and enhancing flow, and don't be afraid to rethink your process or find a new pair of eyes to see the things you can't because you are too close to the issues.

Work at eliminating unplanned work, as if the culture is to keep doing things that are broken, then it's tough to put aside the hours outside of the firefighting to make things work or to innovate. Be continuous, embrace continuous integration and continuous delivery, and keep things going. Make sure that you create cross-functional teams and embrace transparency in the teams and processes. Build mastery and purpose and create single goals everyone can focus on. These are tough things that don't happen overnight, so you need to be consistent in your approach, but also be open to others' needs.

DevOps as a philosophy is about much more than just tools, but the right tools can make the whole process much smoother. Microsoft has released a tool that makes it easier to implement DevOps when working on Azure or with other cloud and on-premises services: Azure DevOps. This is available both as a managed cloud service, as well as a self-hostable product called Azure DevOps Server (previously known as Team Foundation Server, or TFS). To begin using the process, you need to head over to `https://azure.com/devops`. Let's look at the methodology of using Azure DevOps, so you can make better decisions when you are setting it up for the first time. First, Azure DevOps allows you to use four types of project methodology:

- **Basic** is the most fundamental and lightweight of the methodologies. This uses issues, epics, and tasks to track project work. *Figure 3* shows how the association works in the basic template:

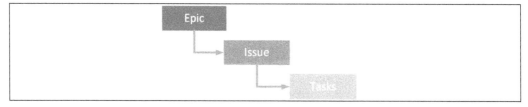

Figure 3: Basic work item process flow

- **Scrum** is lightweight process management. It has more flow than Basic because you use epics, features, product backlog items, tasks, impediments, and bug tasks to track project work. *Figure 4* shows how the association works in the Scrum template:

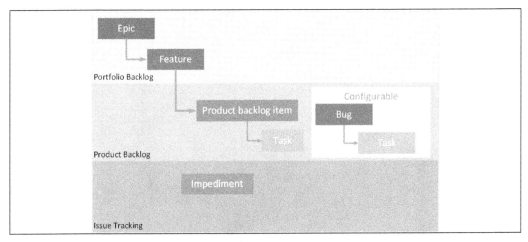

Figure 4: Scrum work item process flow

- **Agile** allows you to work through storyboards and has more process items than Scrum. You have a portfolio backlog for your epics and features, with a backlog for user stories with their tasks and bugs, and there's also an issue tracking system for your projects. *Figure 5* shows how the association works in the Agile template:

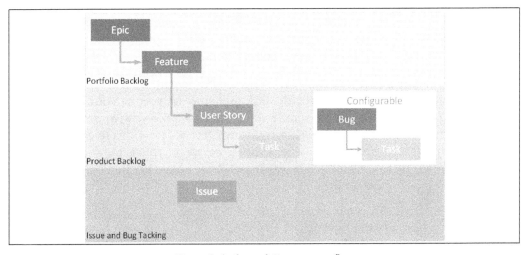

Figure 5: Agile work item process flow

- **CMMI**, which stands for **Capability Maturity Model Integration**, is the most formal project method. It helps maintain an auditable record of decisions. This process helps you track requirements, change management, risk, and reviews. It has a portfolio backlog for your epics and features, which tracks your requirements and their tasks and bugs in your backlog while tracking change requests, issues, risks, and reviews in the issue, change, and risk management project section. *Figure 6* shows how the association works in the CMMI template:

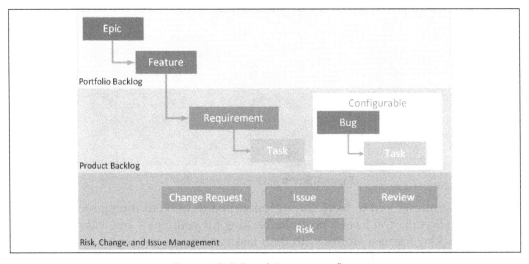

Figure 6: CMMI work item process flow

As you can see, there is a wide range of methodologies you can use to help manage your projects and project portfolios. You can use Azure DevOps as the source of truth for all of your organization's projects, whether they require code or not. You can create a definition around what task completion means and make it consistent across projects, so the team knows the definition regardless of the project.

With these processes, you can either fully automate the delivery of your software all the way to production or set up semi-automated processes with approvals and on-demand deployments for any platform. Azure DevOps provides five main areas that help support these processes in one software package. This is the same for every methodology you pick. Let's look at how each of these areas functions:

Azure Boards	Azure Repos	Azure Pipelines
This helps to manage requirements for your projects through backlogs with dashboards and integrated reports, and helps to quickly track backlog items, user stories, tasks, features, and bugs.	This hosts code repositories. It currently has two flavors, Git and **Team Foundation Version Control (TFVC)**. It helps with managing versions and backing up your application source code.	This is used to build and test your code projects and make them available to other processes. It combines CI and CD with continuous testing to keep your code shippable.
Azure Artifacts	Azure Test Plans	
This is the concept of multiple feeds that you use to control access and organize your packages. This is how you make your package available to other projects, like nuget.org.	This helps track and manage your manual and exploratory testing techniques.	

As you can see, Azure DevOps provides a rich set of tools to help any project to complete, and provides an extension for integration into third-party applications, such as Slack or Trello. One of the biggest additions was increased integration with GitHub for a more collaborative, open-minded, and open source software development platform that Microsoft purchased. Azure DevOps allows you to connect your boards to github.com to manage commits and push requests while linking them to work items within the system, also allowing issues to be tracked with GitHub. This gives you the ability to use Azure DevOps boards to manage your GitHub project. Now let's dig into how they can be used to deliver applications in the cloud.

In the following scenarios, we'll take a deep dive into setting up Azure DevOps. We'll run through the author's personal workflows and opinions and how they can be applied to your own projects.

Capturing requirements in Azure Boards

I like to start all my projects by capturing the requirements in the backlog and using Azure Boards to manage them. Before we get started, you will need to configure things based on the project structure template that you intend to use. I generally choose Scrum, so I will show you some things you need to configure before creating a backlog for the Scrum template.

I use Scrum because it has some formal processes, such as daily updates and planning, and it has a bit more interactive control so you can course-correct when needed. You should start with paths and iterations, which are stored under **Project settings**, as you can see in *Figure 7*:

Figure 7: Azure DevOps paths and iterations

Paths are used to group work items together based on production, business area, or features. Iterations are used to break up work time into management time segments, such as every two weeks in a scrum. Once this is completed, you enter your backlog items so your project can begin.

I like to create a definition of "done" for each task to help ensure a standard measuring stick for work items or tasks to be completed. When you tell someone how their work is measured, you will get better results:

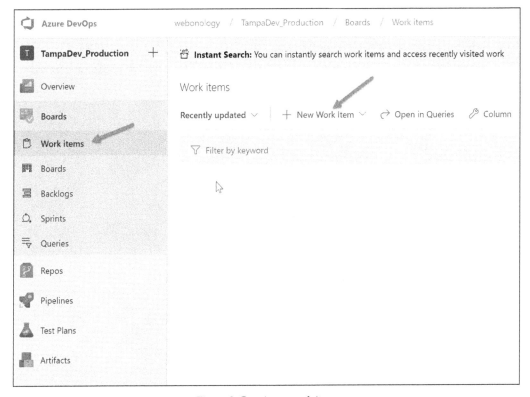

Figure 8: Creating a work item

Once you have a backlog, you can work with the team members to size and break them up into chunks or tasks. Once this is done, the team members will pull items from the backlog based on their capacity for that project. Each team member responsible for a chunk of code should check that code against the work item the code is for. For example, if I am creating a new database element using code first, I would check the code for any changes against the work item I was using to get the requirements.

I always create an enhancement or wish list area for those things that come in as the product moves toward completion that aren't part of the requirements of a minimum viable product. Managing your backlog and tracking requirements is really what works for an organization, so I would research and use the template that's closest to the management style that works for you and the project. Remember that each project can be managed differently. Now let's cover the methodology relating to building and deploying.

Build, deploy, manage

Azure Pipelines handles the builds and deployments of your application code. You generally want to start by creating a code repository, and then you will create a CI build. This is an automation process that builds the code and runs a test on your code every time a developer does a check-in. As you can see in *Figure 9*, **Builds** and **Releases** can be found under **Pipelines**:

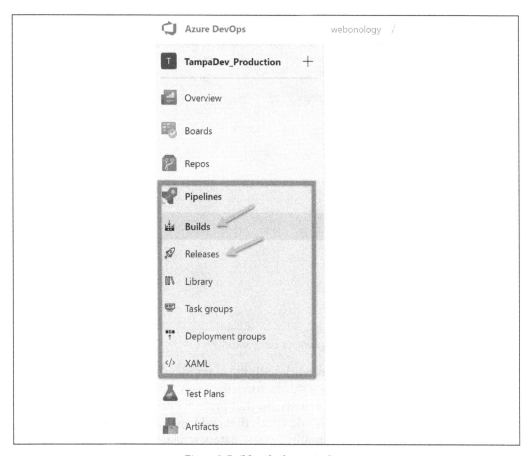

Figure 9: Build and release pipelines

You can then create a release process that uses the build artifacts and deploys them to an environment such as Dev Integration or QA for further system testing. For my build processes, I tend to use a NuGet process for my shared resources to help with reuse and version change rollouts. With that, let's go into CI/CD in a little more depth.

Using CI/CD for high productivity development

CI and CD pipelines help automate steps for builds, tests, and releases with quick feedback while keeping your application in a shippable state. This type of process helps reduce manual errors and helps configure the source of truth for configurations, as CI is predicated on code merges and changes in a central code repository that has no limit on check-ins. We should understand how this affects code variations that other developers need to merge with their repositories, which usually trigger a build-and-test as well as a release in most cases.

It is good to understand the stages of this process, which are source, build, test, and release. The source stage is how developers manage the code base and gated check-ins allowed for builds. Those builds are the second part of these stages because code should build with new changes. We then run tests to help with code quality and meeting requirements. The last part is the release, which is moving that code out into the wild. The CD part of the process is the delivery: being able to deliver an application regardless of the state of the code. With this, we need to understand some high-level best practices.

Implementing DevOps best practices

To implement Azure DevOps best practices, you should remember the following:

- Encourage a DevOps culture by defining shared goals that span the Dev and Ops teams
- Make sure your workflows are flexible; give the team the ability to choose the tools that highlight their expertise
- Adopt more automation and look for opportunities to increase automation to help with continuous delivery
- Encourage small project footprints by leveraging microservices and containers
- Try not to customize security or security groups
- Create a sustainable code review process
- Engage this process early and start with automation from the get-go
- Be continuous
- Communicate

Expediting your application life cycle management process

We discussed the pillars of quality for software development earlier in the chapter. So here, let's look at the steps to use in building your life cycle management in Azure DevOps. Making the project small and easy to deploy with its own **Azure Resource Manager** (**ARM**) template for resources will help with rapid deployment and application independence. Here is one approach:

1. Gather customer requirements
2. Use those requirements to build a backlog and create a definition of "done"
3. The team starts sizing the backlog and breaking out tasks into smaller manageable chunks, as well as building test plans to test requirements
4. Build your interactions and paths so your sprints can be defined as tasks
5. Use a Git repository and build a pull request process to test check-in quality
6. Trigger build and release to the dev integration environment
7. Test the build and verify it
8. Send an email to QA for release approval of cleared code
9. Trigger the release after approval and run the build verification tests on the QA environment
10. Release to production based on the release process to a staging slot, then swap the slots to a new slot or run tests in production cutover if needed

So, with this understanding, let's look at how stages and environments in Azure DevOps work in our release pipelines.

Understanding stages and environments

Azure DevOps has the concept of stages and environments, which are basically deviations within a pipeline based on an element such as tasks to complete (stages) or places they end up (environments). A stage is a way to organize your pipeline for major divisions in the pipeline. They are pieces of the pipeline that can be paused and can perform various checks. The stage can contain multiple stages or parallel stages in the completion of the pipeline.

Environments are the resource endpoints, meaning where or what the artifacts are deployed to. You can have multiple environments, as well as processes such as approvals before release. Now that you have a quick understanding of stages and environments, let's look at deploying and managing services.

Deploying and managing services

First, it is important to understand that resources in Azure are "deployed." This is the provisioning process for Azure, but it is not typically referred to as provisioning. With Azure, there was a transition from a developmental approach to the deployment of resources, which is a JSON representation of the resource that is being deployed. This is referred to as **Infrastructure as Code** or **IaC** and is handled through a system called **Azure Resource Manager** (**ARM**) templates.

ARM templates are the deployment and management service for Azure. They are the management layer that provides the ability to create, update, and delete resources, as well as the desired state of the resource's configuration or DSC. The ARM template system can be used through the portal, PowerShell, Azure CLI, Rest APIs, or Client SDKs, as you can see in *Figure 10*:

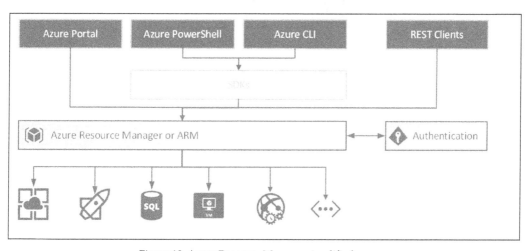

Figure 10: Azure Resource Manager simplified process

As you can see, the deployment system can use either an ARM template or a PowerShell/CLI process to deploy resources. It is important to understand that the PowerShell/CLI system can use ARM templates or deploy resources without an ARM template by directly using the ARM APIs in Azure. It is recommended that you use ARM templates as much as you can to control your deployments. Before we dive into ARM templates or PowerShell/CLI, let's discuss some terminology. The first thing to understand is that all deployments happen to a resource group, which is what I use for application boundaries, and you want your deployment to be small and easy to manage. The things you deploy, such as a VM, an Azure SQL server (or database), or a web app, are the resources that you deploy within the resource groups. Each resource has a provider, called a resource provider, which tells the deployment the type of resource being provided. So, with that, let's dive in a little deeper.

Understanding ARM templates to deploy artifacts

Before we get started on this section, make sure that you have installed the Azure development workload in Visual Studio, because the ARM template project will not be available until you do. In your search bar, type `Visual Studio Installer`, which will give you the screen shown in *Figure 11*:

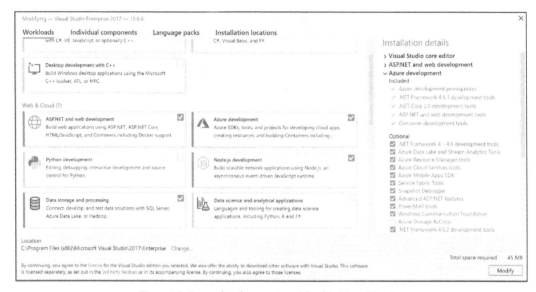

Figure 11: Azure development workload in Visual Studio

Ensure you have the **Azure development** workload checked, as you can see in *Figure 12*:

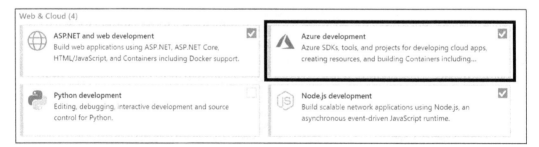

Figure 12: Azure workload

Once you install the workload, you will be able to create an **Azure Resource Group** template within Visual Studio, as you can see in *Figure 13*:

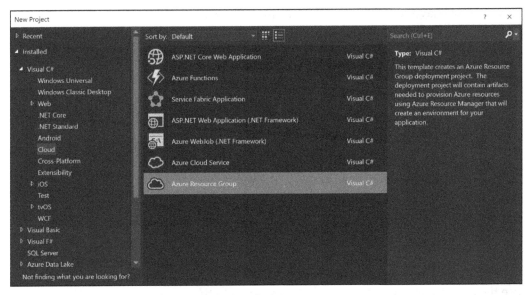

Figure 13: Visual Studio – Azure resource group

Now that we have ensured that Visual Studio is set up to use Azure resource groups, we can dive into how to use them. The ARM templates are the JSON files used to deploy the resources to a resource group. They use a declarative syntax that uses the base elements we discussed. They span four levels within Azure: they can be used in blueprints in management groups, contained in libraries in subscriptions, contained in deployment histories (export scripts) in resource groups, or directly to the resources, as you can see in *Figure 14*:

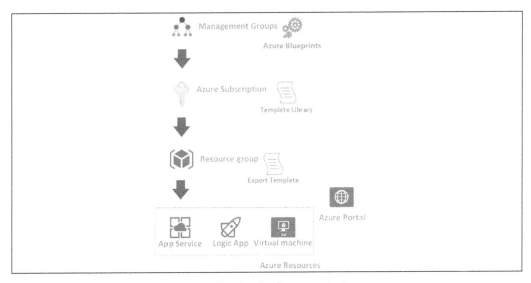

Figure 14: Four levels of resources in Azure

Now that we have set a basic understanding of how an ARM template's scope can be used, let's look at what makes up the basic structure of the template system used to deploy resources to Azure. This is the basic makeup of the JSON file created to house all the resources you need for your project:

```
{
    "$schema": "http://schema.management.azure.com/schemas/2015-01-
    01/deploymentTemplate.json#",  "contentVersion": "",
    "parameters": { },
    "variables": { },
    "functions": { },
    "resources": [ ],
    "outputs": { }
}
```

Let's look at what these elements mean and gain a better understanding of how to use these elements, and also see what they mean to the schema:

Schema	JSON Schema Location	Required?
contentVersion	Version tracking value, default is 1.0.0.0	Yes
parameters	Values that are provided from an external source such as parameter files and parameters from a deployment pipeline	No
variables	Common values provided for deployment	No
functions	User-defined functions	No
resources	Type of resources you are deploying	Yes
output	Return values	No

Important note:

ARMVIZ (http://armviz.io/) is a useful tool for visualizing your ARM templates.

I prefer to use ARM templates as much as I can to deploy my resources, but there are occasions where I cannot use ARM and need to use something else, such as PowerShell/CLI. It is also helpful because you can use Cloud Shell to run the PowerShell/CLI from the library you create, which I like using because I can access Cloud Shell from anywhere within a browser. I use it to execute runbooks and ARM templates, and to change user security from my phone or device, which means I can do more without my computer. Let's look at using PowerShell for deployment.

Using PowerShell to deploy artifacts

All resources in Azure can be deployed using PowerShell, not just ARM templates, and the declarative nature of the template is used to run the PowerShell to create the resources. There are also resources that aren't deployed using ARM, such as DNS or Azure App Services Certificates. These can also be run in Azure DevOps Pipelines, as you can see in *Figure 15*:

Figure 15: Pipeline sample

Azure DevOps powerful security and retention measures for workloads

Azure DevOps provides security out of the box, as well as retention policies that can be customized if you need to. I try to stay away from too much customization around security in Azure DevOps, but let's review security and retention policies to get a better understanding of how they are applied by default.

Security

Azure DevOps is a safe, available, secure, and private delivery tool for organizations. The main things to understand is how authentication, authorization, security groups and roles, permission, and access levels work within the Azure DevOps platform. Authentication can be integrated with ease to **Azure Active Directory** (**Azure AD**), **Microsoft Account** (**MSA**), or on-premises AD, with Azure AD and MSA that are cloud-based solutions. They adhere to **two-factor authentication** (**2FA**), which is a two-level access process using either a mobile device or an emailed code in order to access resources.

While this will get you into the platform, you will need to be assigned a license (basic), be a stakeholder (a business person with no code check-ins), or a Visual Studio subscriber (a development platform user) to see the initial landing page of Azure DevOps. Now that you have logged into the platform, let's look at how authentication methods work for other services within the Azure DevOps platform.

Once access to the platform is created, you will need access to your code repositories if you are a developer and the integration services to Azure DevOps resources. If you don't want to log in each time to the resources that you are using, you can use one of the following things:

- Personal access tokens, which allow you to access a resource from clients such as XCode or NuGet, which don't really support Azure AD credentials

- OAuth, which enables access to the REST APIs of Azure DevOps, and it should be pointed out that only the Profile and Account APIs support OAuth

- SSH authentication, which is the authentication key for Linux, macOS, or Windows platforms running Git for Windows

I should point out that, by default, your collection and account allow access for all the authentication methods listed previously. Authorization within Azure DevOps is based on the individual user or group assignment, which enables the necessary permissions to the service, function, object, or method. Azure DevOps has some preconfigured security groups that you will interact with, but let's look at how permissions are assigned to groups, levels, and states:

Security groups	Permission levels	Permission states
They are assigned at the project level, organization or collection level, or server level for Azure DevOps Server, which is the on-premises server.	These are assigned at the object level, project level, organization or collection level, or server level for Azure DevOps Server, which is the on-premises server.	For permission states, a user or group can have permission to **Allow** or **Inherited Allow** (from the parent object) or to **Deny**, **Inherit Deny**, or **Not Set** for permissions on a task.

Now that we have a brief understanding of the setting permissions, let's look at the built-in groups and what level, server level, collection level, project level, or object level, they sit at.

These are the server-level groups in Azure DevOps Server (on-premises only):

- Team Foundation Administrators: these are like the global admins
- Project Server Integration Service account
- SharePoint Web Application Services
- Team Foundation Proxy Service account
- Team Foundation Service account

These are the collection-level groups:

- Contributors
- Project Collection Administrators, the global admin for collections
- Project Collection Build Administrators
- Project Collection Build Service accounts
- Project Collection Proxy Service accounts
- Project Collection Service accounts
- Project Collection Test Service accounts
- Project Collection Valid Users
- Security Service Group

These are the project-level and object-level groups:

- Build Administrators
- Contributors
- Readers, the lowest level of access
- Project Administrators, global admin for project or object level
- Project Valid User
- Release Administrator
- `<Team Name>` created for the team in the project by default

Azure DevOps centers around three functional access control areas for management. They are Membership, Permission, and Access Level. Membership is assigning the user account to the default group or security groups for access to the project.

Permission management is controlling access to specific functional tasks at different levels of the system, such as object-level permission set on a pipeline or folder. The access level is at the portal level and is defined as **Basic**, **Stakeholder**, or **Visual Studio Enterprise**.

As you can see, there is a robust security element to Azure DevOps. You can find these settings in the Azure DevOps portal under **Project settings | General | Security**, as shown in *Figure 16*:

Figure 16: Azure DevOps project security

Now that we have seen how security works out of the box for Azure DevOps, let's review how retention works within your projects.

Retention

Always ensure your retention policies are configured properly to support your organization's auditing needs within Azure DevOps.

These policies apply to pipelines and test artifacts to retain the history of things that have happened and the artifacts created, as you can see in *Figure 17*:

Figure 17: Azure DevOps retention policies

Summary

As you can see, DevOps and cloud adoption go hand-in-hand, as you want to automate your application deployments in the cloud as much as possible. Azure DevOps provides a great toolset out of the box with Azure Boards, Azure Repos, Azure Pipelines, Azure Artifacts, and Azure Test Plans. Azure DevOps allows you to pick some standard approaches to process delivery with frameworks such as Agile and Scrum. Remember that a proper DevOps practice is everyone's responsibility and isn't about how much stuff you can get done, but on how stuff gets done. In the next chapter, we'll look at the business benefits that arise from moving to the Azure platform.

4
Optimization and Management in Azure

Introduction

Cloud architectures with virtualization create new challenges for management and optimization. Instead of a traditional IT approach of dedicating specific computing resources to specific applications and overprovisioning to ensure enough compute power or storage, the cloud shares resources dynamically. This opens the door to flexibility, scalability, and infrastructure efficiency. However, it also increases the complexity of monitoring, tuning, identifying and resolving issues, and maximizing cost-effectiveness.

In this chapter, we look at management, first in Azure-only resources, then in hybrid environments that can leverage Azure management tools. We continue with a discussion about diagnosing service problems in Azure and how to get support. Finally, because the cloud and Azure specifically must also deliver on the promises of affordability compared to other architectures, we finish the chapter by presenting Azure opportunities for cost savings and optimization.

Managing and optimizing your Azure resources

When you choose Azure, you open the door to many different cloud opportunities. Scalable, pay-as-you-go **virtual machines** (**VMs**), geo-localized data storage, websites, queue management, and artificial intelligence are just some of Azure's capabilities. The control plane of Azure provides multiple tools to help you do this.

Azure Resource Manager

As a component of the Azure control pane, Azure Resource Manager is the deployment and management service for Azure. It lets you create, update, and delete resources in your own Azure space. Examples of Azure resources are VMs, storage accounts, web apps, databases, and virtual networks. As Azure resources proliferate, you can save time and effort by applying management actions across groups of resources, as well as to individual ones. These group-oriented actions can have multiple advantages:

- Prevention of security issues such as connections from multiple locations or from suspicious IP addresses by applying access control to all services in a group

- Logical organization of all the resources in a user subscription by applying tags to those resources

- Deployment of resources in the correct sequence by defining the dependencies between them

- Clarification of the billing at an organizational level by viewing costs for a group of resources, each with the same tag

You can also use declarative templates that specify what you want to happen (the result) to resources, instead of spelling out each individual step or command to detail how you want those things to happen (the method). Thus, for example, you can define a declarative template to immediately create a storage account for VM disks faster, more easily, and more reliably than trying to list all the necessary steps to make and configure the account.

Azure provides four levels of scope: management groups, subscriptions, resource groups, and resources. You can apply management settings at any of these levels of granularity to determine how broad or focused the effect is. Lower levels inherit settings applied at higher levels.

In addition, governance functions in Azure show how IT projects contribute to meeting business objectives. These functions help your organization achieve its goals through effective and efficient use of IT.

Azure Automation

Azure Automation provides a cloud-based automation and configuration service for consistent management across Azure and non-Azure environments such as on-premises datacenters. It comprises process automation, update management, and configuration features. Azure Automation provides complete control during deployment, operations, and decommissioning of workloads and resources.

You can use Azure Automation to automate manual, long-running, error-prone, and repetitive tasks that are commonly performed in a cloud and enterprise environment. This automation lets you save time. It also increases the reliability of regular administrative tasks with the option of automatically performing them at regular intervals. You can automate processes using runbooks, or automate configuration management using **Desired State Configuration** (**DSC**) (see the following section).

Some common scenarios for automation are as follows:

- **Build/deploy resources**: Deploy VMs across a hybrid environment using runbooks and Azure Resource Manager templates, with possible integration into development tools such as Jenkins and Azure DevOps

- **Configure VMs**: Assess and configure Windows and Linux machines as required for the infrastructure and application

- **Monitor**: Identify changes in machines that may cause problems, then remediate or escalate to management systems

- **Protect**: Quarantine a VM if a security alert is raised, and set in-guest requirements

- **Govern**: Implement role-based access control for teams, and recover unused resources

Configuration management

Azure Automation DSC is a cloud-based solution for PowerShell DSC that provides services for enterprise environments. You can manage your DSC resources in Azure Automation and apply configurations to virtual or physical machines from a DSC Pull Server in the Azure cloud. The solution offers a range of reports, including reports that inform you of important events such as nodes moving into non-compliance. You can monitor and automatically update machine configuration across physical and virtual machines, both Windows and Linux, in the cloud and on-premises.

The Automation gallery contains runbooks and modules that accelerate the integration and authoring of processes from the PowerShell gallery and Microsoft Script Center.

Storage management

Azure Storage Explorer lets users manage the contents of their storage accounts. With this standalone application, you can upload, download, and manage blobs, files, queues, tables, and Cosmos DB entities, and manage VM disks.

You can use Azure Storage Explorer to work with Azure Resource Manager or classic storage accounts, to work with Azure Storage data on Windows, macOS, and Linux, and to manage and configure **cross-origin resource sharing** (**CORS**) rules. Data lake management is a further possibility.

Azure Storage Explorer provides several options for connection to storage accounts, including:

- Connection to storage accounts associated with your Azure subscriptions
- Connection to storage accounts and services shared from other Azure subscriptions
- Connection to and management of local storage by using the Azure Storage Emulator

Using Azure Storage Explorer with Azure File Storage

Azure File Storage provides file shares in the cloud using the standard **Server Message Block** (**SMB**) protocol (both SMB 2.1 and SMB 3.0). The Azure File Storage service enables the quick and cost-effective migration to Azure of legacy applications that rely on file shares. With Azure File Storage, you can make data available publicly or store it privately.

Delegated access to resources in a storage account is possible by using a **shared access signature** (**SAS**). With a SAS and without needing to share account access keys, you can give permission to a client for access to specific objects in your storage account for a specified period.

Azure Data Studio

Azure Data Studio is a tool for the management of SQL Server databases, including Azure SQL Database and Azure SQL Data Warehouse systems. Previously called SQL Operations Studio, Azure Data Studio provides an up-to-date editor experience with IntelliSense, code snippets, source control integration (Git), an integrated terminal, and built-in charting of query result sets and customizable dashboards. In Azure, code snippets can generate the correct SQL syntax to create databases, tables, views, stored procedures, users, logins, roles, and so on, and to update existing database objects.

Extending management beyond Azure

Through tools and technology for management, automation, and governance, Azure provides solutions for users to create new applications and recreate existing environments in the cloud to the same levels of governance and regulatory compliance as on-premises deployments. In addition, users benefit from the agility of the cloud in expanding and redeploying compute and storage resources, while Azure resource management tools help them remain efficient, effective, and compliant.

However, users often want to combine Azure and other environments in a hybrid cloud solution that suits their needs better. In the next section, we see how Azure management tools can encompass or be leveraged by non-Azure resources, including automation for monitoring, updating, and other tasks.

Working with your hybrid cloud strategy

Hybrid cloud configurations can offer the best of both worlds, combining elements such as cloud scalability and cost savings with on-premises control and security. It makes sense to combine cloud and on-premises resource management to avoid administration silos and potential waste, while ensuring suitable protection. You can achieve continuous productivity and efficiency through resilient solutions for centralized management, encompassing cloud and on-premises systems. Azure cloud and on-premises Windows servers can be managed in this way using tools such as Windows Admin Center and Hybrid Runbook Worker.

Using local and hybrid management services with Windows Admin Center

Windows Admin Center is a browser-based management toolset for on-premises Windows servers, with access to Azure services. While it can be used to manage Windows servers on private networks that are not connected to the internet, it also has multiple points of integration with Azure services such as Azure Active Directory, Azure Backup, and Azure Site Recovery. Windows Admin Center thus simplifies and extends the management of Windows systems across on-premises and Azure environments. It also manages different generations of Windows Server and Windows 10 systems, and more via the Windows Admin Center gateway installed on Windows Server or Windows 10.

Windows Admin Center contains many tools that should already be familiar to users managing Windows servers and clients. It works with solutions such as System Center and Azure management and security to accomplish single machine management actions. The gateway can be made available through a company firewall for secure management of these resources over the internet using Microsoft Edge or Google Chrome. Role-based access control gives fine-grained control over which administrators can access which management features. Local groups and local domain-based Active Directory are options for gateway authentication, as is cloud-based Azure Active Directory.

Azure hybrid services are available to Windows servers as individual physical servers and VMs, as well as clusters. On-premises Windows Server deployments can then benefit from cloud services that include:

- **Azure Site Recovery** for the protection of VMs with cloud-based disaster recovery. Workloads running on VMs can be replicated to protect business-critical operations if disaster strikes. Windows Admin Center facilitates the setup and replication of VMs on Hyper-V servers and clusters for increased resiliency through the disaster recovery service of Azure Site Recovery.

- **Azure Monitor** with advanced analytics and machine learning for tracking events across applications, infrastructure, and networks. With Azure Monitor, users can monitor server status, events, and performance; set up email alerts; and see apps, services, and systems connected to any server.

- **Azure Network Adapter** for easy Azure network connectivity. On-premises servers can connect securely to an Azure Virtual Network via an Azure Network Adapter.

- **Azure Update Management** for keeping VMs up to date. This tool enables update and patch management for different servers and VMs from one location, whether these servers or VMs are on-premises, on Azure, or hosted by other cloud service providers. Users can rapidly check on available updates, plan update installations, and review and verify update installation success.

- **Azure Backup** for backing up Windows Server. To guard against damage through data accidents, corruption, and attacks, users can back up their Window Server machines and VMs to Azure.

- **Azure File Sync** for syncing on-premise files servers with the cloud. File syncing can obviate the need for server backup on-premises. Files can also be synced across multiple servers by using multi-site sync.

- **Azure Active Directory** (**Azure AD**) authentication for an additional layer of security for Windows Admin Center. Azure AD authentication also enables access to Azure AD's security features, such as conditional access and multi-factor authentication.

These services offer the additional advantages of simple setup and a server-centric view for administrators, being directly integrated into Windows Admin Center. The Azure hybrid services tool in Windows Admin Center brings the integrated Azure services together into a centralized hub to facilitate service discovery in both on-premise and hybrid environments.

When users connect with the Azure hybrid services tool to a server with Azure services already enabled, they benefit from a centralized admin experience for all the services enabled on a server. Users can quickly access the tool required in the Windows Admin Center toolset, connect to the Azure portal for more extensive management of those Azure services, and consult the online documentation.

Windows Admin Center also allows management of Azure VMs, not just on-premises servers. Users can manage VMs in Azure by connecting their Windows Admin Center gateway to their Azure VNet. They can then use the simplified tools of Windows Admin Center.

While Windows Admin Center meets many common needs, it is not designed to replace all legacy **Microsoft Management Console** (**MMC**) tools. For example, Windows Admin Center is complementary to **Remote Server Administration Tools** (**RSAT**), at least until all RSAT management capabilities are surfaced in Windows Admin Center. Similarly, **Windows Admin Center** and **System Center Virtual Machine Manager** (**SCVMM**) are complementary. Although Windows Admin Center can replace MMC snap-ins and recreate a comparable server administration experience, it is not intended to replace SCVMM monitoring capabilities.

Automating resources on-premises and in the cloud by using Hybrid Runbook Worker

With Azure Automation, users can automate manual and repetitive tasks by using runbooks. Using Windows PowerShell or Windows PowerShell Workflow, users can program and deploy automation logic of their choice in these runbooks.

However, runbooks running on the Azure cloud platform may not have access to resources that are on-premise or in other clouds. To extend the reach of runbooks, Azure Automation provides the Hybrid Runbook Worker feature to run runbooks directly on systems that are managing local or other non-Azure resources. The runbooks are stored and managed in Azure Automation, being delivered afterward to the designated systems, which are known as the Hybrid Runbook Workers.

The structure of the runbooks in Azure Automation and in Hybrid Runbook Workers is the same. What differentiates one kind of runbook from the other is that Azure Automation runbooks manage resources in the Azure cloud, while Hybrid Runbook Worker runbooks manage resources local to the Hybrid Runbook Worker or in the environment local to the resources to be automated.

The installation of a runbook is often most simply and reliably done through the automation of the configuration of a Windows system. Manual installation and configuration are also possible. Users with Linux machines can run a Python script to install a runbook agent on their systems.

In Azure Automation, the **RunOn** option allows the specification of a Hybrid Runtime Worker Group. The runbook concerned is then retrieved and deployed by one of the members of this group. If the RunOn option is not used, the runbook is simply run in Azure Automation.

Because runbooks on a Hybrid Runbook Worker access resources outside Azure, their authentication differs from that of runbooks running within and authenticating to Azure resources. Runbooks outside Azure can provide their own authentication to local resources. When running on a local Windows system, they will typically run in the context of the local system account. When the on-premises system is Linux-based, the special user account nxautomation is used.

Runbooks on a Hybrid Runbook Worker can use managed identities when configuring authentication to Azure resources. Alternatively, users can specify a **RunAs** account to create a context for all runbooks. However, using managed identities for Azure resources offers certain advantages over RunAs accounts. Users do not need to export or renew a RunAs certificate that must then be imported into the Hybrid Runbook Worker. They are not required to write their runbook code to handle the runbook connection object either.

Hybrid Runbook Worker for updates and monitoring

By enabling the Update Management solution, you can automatically configure a Windows computer to support runbooks for update management. This applies to Windows computers connected to your Azure Log Analytics space. These computers then become Hybrid Runbook Workers. This allows you to extend your existing infrastructure and apply updates from a reliable, secure, and centralized location, creating a more integrated hybrid cloud/on-premises infrastructure. On the other hand, the Windows computer concerned is not automatically registered with any existing Hybrid Worker Groups in the user's Automation account.

The Microsoft Monitoring Agent can be installed to connect computers with Azure Monitor logs. When the agent is installed on an on-premise computer and connected to the user's workspace, it will automatically download the relevant Hybrid Runbook Worker components. These components include the `HybridRegistration` PowerShell module, which in turn contains the `Add-HybridRunbookWorker` cmdlet. Running this cmdlet installs the runbook environment on the computer and registers it with Azure Automation.

Building on and extending hybrid cloud management possibilities

Automation of processes on-premises or in a non-Azure cloud environment can be copied from or modeled on successful Runbook Worker deployment in Azure. The resulting Hybrid Runbook Workers are only constrained by the limits of the resources on the Hybrid Runbook Worker itself. Thus, the Hybrid Runbook Workers are free from certain limits that are imposed on Azure sandboxes, such as disk space, memory, network sockets, or running time.

Windows Admin Center was also designed to be extensible. Microsoft makes a **software development kit (SDK)** available for Microsoft and third-party developers to create their own Windows Admin Center tools and solutions and build on what is currently available.

What if something goes wrong?

When business or other key workloads run on Azure, users need to know that their Azure resources are available and working properly. Conversely, if there is a problem, they must be alerted. Data must be available to verify that **service level agreements (SLAs)** are being respected. Scheduled maintenance on Azure resources by Azure needs to be communicated and integrated into customers' planning.

In the next section, we consider the wide variety of support tools available for Azure for global, personalized, and individual resources. Resource status, service health, alerts, and integration with Azure Monitor (described earlier in this chapter) are also discussed, as well as the practical aspects of getting support directly from Microsoft.

The next section in this chapter deals with the optimization of budgets and cost savings, including analysis and management of costs and Azure solutions to help you make your cloud spend go further.

Azure cost savings – visibility, accountability, and optimization

Enterprises often look to cloud services as a way of reducing IT costs. However, any cost savings will also depend on how enterprises manage their costs and optimize their cloud spending. As with business activities, different departments will need to collaborate for effective cloud cost management. IT, finance, and different levels of management are all likely to be involved to correctly analyze costs, control them, and prepare future budgets as needed.

Azure offers a range of different tools to help enterprises manage their costs. A pragmatic business approach and common sense are also important. Finance departments should understand where cloud costs are generated and how cloud spend is trending, but so too should cloud IT teams.

1. **Visibility**

   Cost analysis can help users and stakeholders to explore and break down cloud costs. Cost aggregation can show them where the largest amount of funds is being spent and what to expect in the future. Information on accumulated costs over time can let them track costs by month, quarter, or year against budgets.

2. **Accountability**

   Identifying specific entities that must fund the use of given resources encourages efficiency and cost-effectiveness, as well as helping to avoid unwanted surprises in billing.

3. **Optimization**

   Cost analysis can help organizations plan for better resource usage (right-sizing), highlight wasted resources, and improve plans for cost estimations.

Azure Cost Management

Azure Cost Management helps you to plan cloud resource consumption while paying attention to costs. It enables you to analyze costs effectively and optimize your cloud spend. With Azure Cost Management, you can see trends and patterns in usage and cost for your organization, with analytics to explore this data. It makes reports available to you on usage-based costs for Azure services and offerings from Marketplace third-party offerings.

The cost data for the reports takes current organizational prices and other Azure discounts into account. The reports also help you to assess possible spending anomalies, using Azure management groups, budgets, and recommendations to help you see possible opportunities for cost reduction. You can use the predictive analytics of Azure Cost Management to manage or plan for costs into the future. In addition, the Azure portal and Azure APIs let you automatically export and integrate cost data with other systems and processes with periodic reports.

Starting to optimize your cloud investment

Azure tools and a methodical approach to cost management can help you optimize your cloud spend. While Azure already facilitates the construction and deployment of cloud solutions, it is still important to ensure that those solutions are tuned for cost-effectiveness.

Good cost management starts before budgets are used. It depends on having suitable tools, assigning accountability for costs, and optimizing expenditure. These principles should be understood and can be applied by at least three groups.

- **IT teams** that manage cloud resources daily must adjust their activities in a timely way to generate the most value to the business for the budgets they use for those cloud resources

- The **finance department** must check requests for budget against financial goals and forecasts of cloud expenditure

- Finally, **managers** must ensure that cloud spending and results are in line with the business goals of the organization

Using scopes for Azure cost management

Most Azure resources are deployed in resource groups, which are part of subscriptions. Authorized Azure users view and manage the cost aspects of Azure resources via scopes, which are nodes in the Azure resource hierarchy. For cost management, Microsoft defines two roles:

- Billing data management (invoices and payments, for example)

- Cloud services management (governance of policy and costs, for example)

Azure resource management is also done via scopes but uses **Azure Role-Based Access Control** (**RBAC**). The two types of scopes are called billing scopes and RBAC scopes to differentiate them. RBAC scopes do not differ according to the Azure subscription type, but billing scopes may vary to extend from single resource groups to entire billing accounts.

The following Azure scopes apply per subscription for cost management by user and group:

- **Owner**: Authorized to create, change, or delete subscription budgets
- **Contributor and Cost Management contributor**: Authorized to create, change, or delete their own budgets, and change the budget amount for budgets of others
- **Reader and Cost Management reader**: Authorized to view specific budgets

Cost management life cycle

Cost management is an iterative process encompassing the four activities below with several stages that form a loop or virtuous circle. This life cycle should be known and applied by all the people or teams involved in cloud cost management.

Planning

As the saying goes, "Plans are worthless, but planning is everything." Any plan for using cloud resources to meet business goals is only an approximation because circumstances can alter rapidly. However, by continually considering changes in the goals and situation of the enterprise, plans can be updated as often as needed to remain realistic. Key questions to ask regularly and frequently are:

- What business goals and challenges must my enterprise meet?
- How is cloud usage likely to evolve if those goals and challenges are to be met?

The answers to these questions will help you to identify the Azure resources and infrastructure that best suit your enterprise.

Monitoring

As you implement your plan, you need to know how much your enterprise is spending and on which cloud resources. Underused resources should be used better or canceled, waste must be eliminated, and opportunities to save money without impacting business goals must be maximized.

Accountability

Financial accountability is the responsibility for the way the budget is used and managed. This responsibility must be clearly and precisely assigned. Costs incurred can then be attributed to specific projects or departments. Spending efficiency can be monitored effectively.

Overspending can be identified down to the project or resource level and appropriate measures taken to bring costs into line again or to justify new budgets corresponding to the value gained for the enterprise.

Optimization

Spending optimization can be accomplished in two ways. The first is by checking performance, goal achievement, or other relevant results against financial outlay. The smaller the outlay for a given result, the better the spending is optimized. The second way is to leverage purchase and licensing optimization and infrastructure changes (see below).

Analyzing and managing your costs

Once an Azure solution has been implemented, it is important to know how costs vary over time.

Organizing and tagging resources

Tags are an effective solution for financial accountability. They allow you to attribute cost to specific projects or teams, and group costs together for further analysis. Alternatively, Enterprise Agreement customers can define separate subscriptions for departments or projects, which also helps promote accountability and individual efforts for cost reduction. Subscriptions and resource groups can be useful ways of organizing and attributing costs to different parts of your organization.

Analysis of usage cost

Regular analysis of costs compared to usage can be useful for spotting usage trends and monitoring the evolution of costs for specific projects and teams. Key points can include:

- Estimated costs for the month and comparison with the corresponding budget
- Identification of spend anomalies, showing costs that fall outside a reasonable range and any other exceptional costs
- Reconciliation of invoices to highlight any unexpected cost increase or changes in spending trends
- Chargeback to internal consumers with a breakdown of charges per project, department, or other entity

Billing data can be automatically exported into other applications, such as a financial system or a data visualization dashboard. Instead of manually retrieving files, users can configure automated exports to Azure Storage with automatic integrations of the Azure Storage data into other systems.

Budget creation

Good estimates, spending pattern analysis, and forecasts are the ingredients for effective budgeting. With Azure Budgets, you can define budgets according to cost or usage with a wide range of limits and alerts. An action can be automatically triggered when a specific budget threshold is reached. For example, VMs can be shut down or infrastructure can be moved to a different pricing level. As budgets are used (budget burn-down), data can be reviewed, and changes made as needed.

Optimizing costs

Optimization of costs comes from maximizing the efficiency of resources and removing those that are not generating value for your enterprise. This includes resources that have been deployed for a project of a fixed duration and that have not been spun down or canceled after the project finishes. An enterprise simulation or system test run over a weekend, for instance, may require considerable compute and storage resources. However, a test team may assume that the operations team will adjust resource levels on the following Monday, whereas the operations team may be unaware of the weekend exercise. Azure Cost Management can help rectify such situations.

Azure Advisor

This service offers different functionalities, including the identification of VMs that are using CPU or network resources at a low level. When you know which machines these are, you can stop them or resize them according to the cost forecast for keeping them in operation. If reserved instance purchases can help you reduce your costs, Azure Advisor can provide recommendations for such purchases based on the previous 30 days of your VM usage.

VM right-sizing

It is important to select the correct size of VMs for your cloud workloads. VM sizing is an important factor in determining overall Azure cost. The number of VMs required in Azure may also be different to the number deployed in an on-premises datacenter. Individual VM size and overall quantity should therefore be calculated to correspond to the compute requirements for the workloads to be run in Azure.

Azure discounts

Volume discounts either in terms of quantity or usage are a common feature of business agreements. Azure takes the same approach, offering cost savings to customers accordingly.

Azure Reservations

Receive a discount on your Azure services by purchasing Azure Reservations. Cost savings can be significant compared to pay-as-you-go prices for VMs, SQL database compute resources, and additional Azure services. You can improve budgeting with a single upfront payment, making it easy to calculate your investments or you can lower your upfront cash outflow with monthly payment options at no additional cost. You can purchase one-year or three-year term Azure Reservations.

Buying an Azure Reservation can be the most cost-effective choice for customers with VMs, Azure Cosmos DB, or SQL databases that are in operation over long periods. For example, without a reservation and with a requirement for five instances of a service, a customer will pay standard pay-as-you-go rates. But by buying a reservation for those resources, the customer immediately benefits from the reservation discount, thus saving money compared to the pay-as-you-go rates.

Azure Hybrid Benefit

The Azure Hybrid Benefit program offers cost savings if you already have on-premises deployments with Windows Server or SQL Server licenses. The Windows Server benefit means that each license includes the operating system for up to two VMs. Users then pay only the compute costs, with the base compute rate being equal to the Linux rate for VMs. Similarly, an existing SQL Server license can bring significant savings on vCore-based SQL database options, such as SQL Server in Azure VMs and SQL Server Integration Services.

Azure Reserved VM Instances

With **Azure Reserved VM Instances (RIs)**, you can reserve VMs in advance for cost savings when combining Azure RIs with Azure Hybrid Benefit. Further advantages of RIs include the exchange or cancellation of reservations and prioritized compute capacity in Azure regions.

RIs also provide a feature called instance size flexibility, which automatically applies the RI savings to any VM that you use within the same region and within the same Azure RI VM group. Instance size flexibility allows you to meet changing needs and realize applicable cost savings without being locked into a specific VM size.

Automated RI management means that Azure can automatically apply RIs to other VM sizes in the same group and region. Advantages such as these apply to both Windows and Linux VMs in Azure.

Reaping the benefits of cost management

When Azure costs are correctly managed, budgetary limits and goals can be adhered to. Financial accountability can be ensured through correct and timely cost data that allows comparison with financial goals. Insights from those data can help identify poor cost-efficiencies, such as underused resources. They can help identify options for improving efficiencies or changes for cost optimization. Azure enables robust cost management processes with recommendations, actions, and verification that changes made are producing the expected cost benefits.

When Azure costs are correctly managed, budgetary limits and goals can be adhered to. Financial accountability can be ensured through correct and timely cost data that allows comparison with financial goals. Insights from this data can help identify poor cost efficiencies, such as underused resources. They can help identify options that will improve efficiency or suggest changes for cost optimization. Azure enables robust cost management processes with recommendations, actions, and verification that changes made are producing the expected cost benefits.

Diagnosing service problems in Azure and getting support

Azure provides a range of tools to help users monitor and manage Azure service situations. Azure Service Health groups together three services to help users see overall Azure status, customized reports on asset groups that affect customers, and detailed information on individual assets. Issues detected by Azure Service Health tools can trigger alerts via text or voice messages, emails, automated responses using Azure-or user-created runbooks, or actions within Azure or within other preferred resources management applications.

Global level status (Azure status)

Azure status information helps users see at a glance the status or impacts on services they use. This overview of the health of all Azure services is part of Azure Service Health. It can also be consulted by any visitor to the Microsoft public **Azure status** page.

Personalized service status (Azure Service Health)

Naturally, users also want to know about the health status of Azure services and regions, as it applies specifically to their resources. Users can access Azure Service Health in order to view communications about outages, scheduled maintenance actions, and other information relating to health and service impacts, according to the services and resources currently used by that user. Users can set up Service Health alerts to notify them over their preferred communications media when the Azure services and regions they use risk impact from service problems, schedule maintenance, or other changes.

Individual asset status (Azure Resource Health)

As an authorized Azure user, you can obtain information from Azure Resource Health on a specific cloud resource such as an individual VM. Via Azure Monitor, you can set up alerts to warn you of changes in the availability of your cloud resources. The combination of Azure Resource Health and Azure Monitor can provide you with better information on a minute-by-minute basis, allowing you to speedily ascertain if an issue is related to an event on the Azure platform or has been caused by a problem in your own environment.

Azure status updates and history

The **Azure status** page is updated dynamically as changes occur in the health of Azure services. You can also define the rate at which this page is refreshed with new data, with running information on the last time the page was updated. Azure status and service health change information is also available via an RSS feed. The **Azure status history** page shows older events up to 90 days in the past, with preliminary root cause, mitigation, and next steps information.

Overview of Azure Service Health

Azure Service Health gives users a dashboard that they can customize to track the health of their Azure services in the regions where they use them. Users can monitor active events such as ongoing service issues, scheduled maintenance for the short term, or other health advisories that concern them. Users can also use the dashboard to create and manage alerts to proactively warn them of service problems that affect them. Events that become inactive are stored in the health history for a maximum of 90 days.

Azure Service Health events

Azure Service Health monitors three kinds of health events that may affect your resources:

1. **Service issues**

   These are continuing problems that may currently affect your resources. The ST view shows you when the problem started, and the services and regions that are affected. This view also makes available the latest update on actions from Azure to resolve the problem.

   The **Potential Impact** tab displays the resources that you own and that could be affected by the problem. This information is also available as a CSV list for download and sharing.

2. **Scheduled maintenance**

   This is maintenance in the short term that could have an impact on the availability of your resources.

3. **Health advisories**

   These concern changes in Azure services that you should know about, such as the deprecation of Azure features or a usage quota that has been exceeded.

A link for a given issue is available for use in a preferred problem management system. Sharing with others that do not have access to the Azure portal is possible through the download of PDF (and in some cases CSV) files. Users can also pin a personalized health map to their dashboard to show their business-critical subscriptions, regions, and resource types via a filter on **Service Health**.

The **Service Health** page provides links for Microsoft support, including cases where a resource remains in an unsatisfactory state even after an issue has been resolved.

Azure Service Health alert configuration

You can use the integration of Service Health with Azure Monitor to receive email, text message, and webhook notification alerts when changes or incidents affect your resources. To receive these alerts, configure an activity log alert for the service health event of interest to you, then use an action group to route the alert to people who need to know about the alert. An action group is a definition of the actions to be taken if an alert is triggered.

Azure Resource Health

Azure Resource Health reports on the existing and historical health of your resources. These reports enable you to diagnose and obtain support for service issues that impact your Azure resources. Whereas Azure status is a "broad brush" report on service problems affecting Azure users in general, Resource Health offers a personalized dashboard to show you specific resource health. For example, Resource Health makes it easy to check if SLAs have been respected by displaying all the instances of unavailability of your resources due to Azure service issues.

Resource health evaluation

Azure Resource Health uses signals from various Azure services to evaluate the health of a resource. The resource may be a VM, SQL database, web application, or any other instance of an Azure service. If Resource Health finds the resource to be unhealthy, it analyzes more information to find the cause of the problem. It also reports on actions by Microsoft to remedy the problem and suggests actions for the user to fix the problem too.

Resource health status

The health status of a resource may be shown as any of the following:

- **Available**

  The **Available** status indicates that no events affecting the health of the resource have been affected. A **Recently resolved** notification is displayed up to 24 hours after a resource recovers from unscheduled downtime.

- **Unavailable**

  The **Unavailable** status indicates that a continuing platform or non-platform event (refer to the *Platform and non-platform events* section) impacting the health of the resource has been detected by the service.

- **Unknown**

  If Azure Resource Health has not received information about a resource within the last 10 minutes, it displays the status as **Unknown**. This may be an important event for subsequent troubleshooting. The status may change to **Available** after a few minutes, if the resource is operating as expected. Otherwise, problems using the resource may indicate that it is being impacted by an event in the platform.

- **Degraded**

  The **Degraded** status indicates the detection of a performance loss for a resource, even if the resource can still be used. Individual resources have their own criteria for reporting a **Degraded** status.

Platform and non-platform events

Multiple components of the Azure infrastructure trigger platform events, whether as scheduled events or unplanned incidents (an unexpected host reboot, for instance). Azure Resource Health provides additional information about the event and recovery from the event. It also enables users to contact Microsoft Support, with or without an active support agreement.

Non-platform events are caused by users, for instance, halting a VM or reaching the limit for Redis connections to Azure Cache for Redis.

Reporting an incorrect status

If you consider a health status for a resource to be wrong, you can use **Report incorrect health status** to report this to Microsoft. You can also contact Microsoft support from Azure Health Monitor if an Azure issue is impacting you.

Integration with Azure Monitor

Azure Monitor collects monitoring (telemetry) data from different on-premise and Azure sources. It also receives log data from management tools like those in Azure Security Center and Azure Automation and can receive health status information from Azure Service Health. Azure Monitor aggregates and stores the telemetry data in a log data store configured for optimal performance and cost-effectiveness.

Users can analyze data, configure alerts, and obtain end-to-end views of their applications via Azure Monitor. They can leverage insights from machine learning to accelerate the identification and resolution of issues. Azure Monitor supports .NET, Java, Node.js, and other popular languages and frameworks.

Azure Service Health can thus be integrated into the centralized, scalable management system of Azure Monitor that unifies operational telemetry and provides advanced tools for improved availability and performance.

Using Azure Monitor capabilities, Azure Service Health can then also be integrated with DevOps processes and tools such as Azure DevOps, Jira, and PagerDuty, as well as with other user management tool favorites such as Grafana, IBM Radar, InfluxDB, SignalFx, and Splunk.

Getting support from Microsoft

Azure users can create and manage support requests via the Azure portal. For example, click on **?** in the top-right corner and select **New Support Request** to create a support request.

The support request experience has been designed to be streamlined, integrated, and efficient for users. A wizard helps users by simplifying the procedure, maintaining the resource context (no need to switch to a context other than that of the resource), and collecting the key information needed for efficient issue resolution. The key information allows the wizard to route the support request to the most suitable support engineer for the issue, so that issue diagnosis and resolution can begin as soon as possible.

Based on the problem category and type selected by the user, Microsoft can also provide contextual self-help information for users to address their issues immediately by themselves. If the recommended solutions do not remedy the issue, the process continues through to the creation of a support request and its transmission to the Microsoft support team.

RBAC for support requests

Azure RBAC lets you define highly granular management access. The Azure portal at `portal.azure.com` uses this RBAC to authorize different levels or scopes of support request creation and management. For example, scopes may extend to a resource, a resource group, or an entire subscription. Users, groups, and applications can have access via the appropriate RBAC role to the level or scope appropriate for them.

For example, a resource group owner with read permissions at the subscription scope can manage all resources in the resource group. These resources might include VMs, web applications and sites, and subnets. However, this resource group owner cannot create a support request for a VM resource in the resource group. To do this, the resource group owner must first be granted write permission at the subscription scope. Alternatively, the role of the resource group manager could be defined to include specific Microsoft Support authority at the subscription scope.

Support effectiveness

The previous section showed how users can benefit from a range of tools in Azure to monitor and manage Azure service resources. These tools range from global, publicly available information (Azure status), through service-and resource-specific notifications (Azure Service Health and Azure Resource Health), to comprehensive integration and management possibilities via Azure Monitor and other service management systems. Users can receive notifications through a variety of channels and responses to service health alerts can be automated using Azure Automation.

Support is user-friendly and efficient thanks to the use of support links and a support request wizard. At the same time, access to support requests is both flexible and secure using Azure RBAC to authorize user actions at the appropriate levels and within the appropriate scope.

Summary

This chapter has explored the management and optimization of resources relating to Azure from different angles, including availability, efficiency, security, and cost-effectiveness. We also discussed hybrid cloud environments, showing how Azure solutions can be used to enhance the management of resources in non-Azure installations. Azure cost-saving and budgeting optimizations were also addressed. We finished this chapter by presenting monitoring and issue resolution from different points of view, including resource health.

Index

A

geo-redundant storage (GRS) **89**
greenfield Azure environment
  deploying **34**

H

hybrid cloud management possibilities
  building **127**
  extending **127**
hybrid cloud models **3, 4**
hybrid cloud strategy
  working with **123**
Hybrid Runbook Worker
  used, for monitoring **126, 127**
  used, for updates **126, 127**
  used, for automating resources in
    cloud **125, 126**
  used, for automating resources
    on-premises **125, 126**
Hyper-V infrastructure assessment **13**

I

IaaS security
  best practices **84**
  reference link **84**
identity control **18**
identity management **87**
identity management, in AAD
  best practices **87**
incorrect status
  reporting **138**
Infrastructure as a Service (IaaS) **5**
Infrastructure as Code (IaC) **109**
IoT architectures, with Azure
  reference link **78**
IoT Central
  reference link **76**
IoT Ecosystem
  architecting **76, 78**

K

key application architectures **64**

L

local and hybrid management services

used, with Windows Admin Center **123, 125**
locally redundant storage (LRS) **89**
Logic Apps
  about **73**
  actions **73**
  billing, types **71**
  Enterprise Integration Pack **73**
  managed connectors **73**
  triggers **73**
  workflows **73**

M

manageable applications
  design principles, on Azure **84, 85**
managed disks **39**
management groups
  about **22**
  reference link **98**
manual migration
  about **29**
  SQL databases, migrating with bacpac **29, 30**
  VHD disks, migrating **29**
message-first development pattern
  reference link **73**
microservice architecture style
  reference link **67**
microservice ecosystem
  architecting **64-66**
  benefits **67**
  best practices **67**
  characteristics **65**
microservices **6**
Microsoft Account (MSA) **113**
Microsoft Backbone cabling **41**
Microsoft Management Console (MMC) **125**
Microsoft's best-practice
  reference link **62**
migration processes **28**
migration tools **28**
mobile applications
  architecting **74**
  principles **75**
modern web application design
  bounded context **79**
  characteristics **79**
  dependency inversion **79**

www.ingramcontent.com/pod-product-compliance
Lightning Source LLC
Chambersburg PA
CBHW060142060326
40690CB00018B/3952